PRAISE FOR *A GRAI*[illegible]

“Schwarcz’s light touches of humor make th[illegible] information feel accessible and ensure that it’s entertaining. With enough facts to soothe anxious, health-conscious individuals as well as some good tidbits to share, this enlightening collection offers every reader something new to learn and marvel over.”
— *Booklist*

PRAISE FOR *A FEAST OF SCIENCE*

“Huzzah! Dr. Joe does it again! Another masterwork of demarcating non-science from science and more generally nonsense from sense. The world needs his discernment.”
— Dr. Brian Alters, Professor, Chapman University

PRAISE FOR *THE FLY IN THE OINTMENT*

“Joe Schwarcz has done it again. In fact, he has outdone it. This book is every bit as entertaining, informative, and authoritative as his previous celebrated collections, but contains enriched social fiber and 10 percent more attitude per chapter. Whether he’s assessing the legacy of Rachel Carson, coping with penile underachievement in alligators, or revealing the curdling secrets of cheese, Schwarcz never fails to fascinate.”
— Curt Supplee, former science editor, *Washington Post*

PRAISE FOR *DR. JOE AND WHAT YOU DIDN’T KNOW*

“Any science writer can come up with the answers. But only Dr. Joe can turn the world’s most fascinating questions into a compelling journey through the great scientific mysteries of everyday life. *Dr. Joe and What You Didn’t Know* proves yet again that all great science springs from the curiosity of asking the simple question . . . and that Dr. Joe is one of the great science storytellers with both all the questions and answers.”
— Paul Lewis, president and general manager, Discovery Channel

PRAISE FOR *THAT’S THE WAY THE COOKIE CRUMBLES*

“Schwarcz explains science in such a calm, compelling manner, you can’t help but heed his words. How else to explain why I’m now stir-frying cabbage for dinner and seeing its cruciferous cousins — broccoli, cauliflower, and brussels sprouts — in a delicious new light?”
— Cynthia David, *Toronto Star*

PRAISE FOR *RADAR, HULA HOOPS, AND PLAYFUL PIGS*

"It is hard to believe that anyone could be drawn to such a dull and smelly subject as chemistry — until, that is, one picks up Joe Schwarcz's book and is reminded that with every breath and feeling one is experiencing chemistry. Falling in love, we all know, is a matter of the right chemistry. Schwarcz gets his chemistry right, and hooks his readers."
— John C. Polanyi, Nobel Laureate

Also by Dr. Joe Schwarcz

A Grain of Salt: The Science and Pseudoscience of What We Eat

A Feast of Science: Intriguing Morsels from the Science of Everyday Life

Monkeys, Myths, and Molecules: Separating Fact from Fiction, and the Science of Everyday Life

Is That a Fact?: Frauds, Quacks, and the Real Science of Everyday Life

The Right Chemistry: 108 Enlightening, Nutritious, Health-Conscious and Occasionally Bizarre Inquiries into the Science of Everyday Life

Dr. Joe's Health Lab: 164 Amazing Insights into the Science of Medicine, Nutrition and Well-Being

Dr. Joe's Brain Sparks: 179 Inspiring and Enlightening Inquiries into the Science of Everyday Life

Dr. Joe's Science, Sense and Nonsense: 61 Nourishing, Healthy, Bunk-Free Commentaries on the Chemistry That Affects Us All

Brain Fuel: 199 Mind-Expanding Inquiries into the Science of Everyday Life

An Apple a Day: The Myths, Misconceptions and Truths About the Foods We Eat

Let Them Eat Flax: 70 All-New Commentaries on the Science of Everyday Food & Life

The Fly in the Ointment: 70 Fascinating Commentaries on the Science of Everyday Life

Dr. Joe and What You Didn't Know: 177 Fascinating Questions and Answers About the Chemistry of Everyday Life

That's the Way the Cookie Crumbles: 62 All-New Commentaries on the Fascinating Chemistry of Everyday Life

The Genie in the Bottle: 64 All-New Commentaries on the Fascinating Chemistry of Everyday Life

Radar, Hula Hoops, and Playful Pigs: 67 Digestible Commentaries on the Fascinating Chemistry of Everyday Life

SCIENCE GOES VIRAL

Captivating Accounts of Science in Everyday Life

DR. JOE SCHWARCZ

ECW

This book is also available as a Global Certified Accessible™ (GCA) ebook. ECW Press's ebooks are screen reader friendly and are built to meet the needs of those who are unable to read standard print due to blindness, low vision, dyslexia, or a physical disability.

Purchase the print edition and receive the eBook free. For details, go to ecwpress.com/eBook.

Published by ECW Press
665 Gerrard Street East
Toronto, Ontario, Canada M4M 1Y2
416-694-3348 / info@ecwpress.com

Cover design: David A. Gee

LIBRARY AND ARCHIVES CANADA CATALOGUING IN PUBLICATION

Title: Science goes viral : captivating accounts of science in everyday life / Dr. Joe Schwarcz.

Names: Schwarcz, Joe, author.

Identifiers: Canadiana (print) 2021024030X | Canadiana (ebook) 20210240377

ISBN 978-1-77041-650-5 (softcover)
ISBN 978-1-77305-769-9 (ePub)
ISBN 978-1-77305-863-4 (PDF)
ISBN 978-1-77305-864-1 (Kindle)

Subjects: LCSH: Science—Popular works.

Classification: LCC Q162 .S395 2021 | DDC 500—dc23

This book is funded in part by the Government of Canada. *Ce livre est financé en partie par le gouvernement du Canada.* We also acknowledge the support of the Government of Ontario through Ontario Creates.

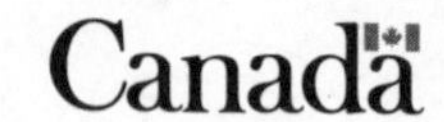

PRINTED AND BOUND IN CANADA

PRINTING: MARQUIS 5 4 3 2 1

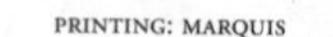

CONTENTS

INTRODUCTION

These are turbulent times. During the past year, terms like pandemic, spike protein, viral particles, variants, mRNA vaccines, viral vectors, antibodies, hydroxychloroquine, social distancing, immune response, convalescent plasma, aerosol transmission, viral load, and of course, face coverings, have become part of "normal" conversation. The truth is that our life these days is anything but normal. I now lecture online, our staff meetings are via Zoom, restaurant meals are a distant memory of the past, my gym is closed, sports are played without fans, theaters sit empty, curfews are in place, travel is out of the question, and I can only see my daughters and grandchildren on FaceTime. Thank goodness for that and for Netflix!

Ask someone to describe what the world has gone through since January 2020 when we first heard about some strange cases of pneumonia in Wuhan, China, and we are likely to hear "surreal," "unbelievable," "mind-boggling," "unthinkable," or "unimaginable." As we were cruising through our more or less contented lives, who could have imagined that such a curse would befall the world, upsetting every proverbial apple cart in existence? Actually, there were scientists who not only imagined it but were certain that such a global scourge was inevitable. After all, there had been plenty of plagues, epidemics, and pandemics in the past, and there was no reason to think that the world had become immune to such calamities. Indeed, the film *Contagion*, produced in 2012 with consultants from the Centers for Disease Control and Prevention (CDC), portrays a fictional pandemic that eerily resembles our current reality.

When I first started thinking about what to include in this book, I had never heard of Wuhan. I thought I would feature my usual array of diverse and sometimes esoteric topics like intermittent fasting, placenta creams, biblical dyes, essential oils, Jean Harlow's hair, Lincoln's magician, bioplastics, along with assorted examples of quackery. Then the cursed plague hit. A giant elephant in the room! It could not be ignored. I had to address it, and I have. With difficulty. That's because the situation is dynamic, and what was true yesterday is not true today, and what seems to be true today may not be true tomorrow. That calls for an apology in advance for sometimes being out of date by the time you read some of the COVID-related accounts, an inevitability in this crazy situation.

I also ran into another stumbling block. As I began to put this volume together in the summer of 2020, the horrific murder of George Floyd took the world by storm. Because of the emergence of COVID-19, and its tentacles ensnaring every facet of our lives, it already seemed frivolous to discuss elderberry extracts, the chemistry of Lego, or plant-based burgers. But how can one write about hair dyes or bisphenol A in socks after witnessing the last breath being squeezed out of a man in the most hideous fashion? How can one talk about hydrogen peroxide oxidizing para-phenylenediamine when millions are at the mercy of a society speckled with hatred and racism? You can't.

So before launching into discussions of COVID-19, I feel a need to have a say about another plague that also threatens to unravel the fabric of our lives, racism. I am no expert on the subject, but being the son of a Holocaust survivor, I am very sensitive to prejudice, intolerance, and xenophobia of any kind. My mother was deported to Auschwitz in 1945, where she was actually saved from the gas chamber by Dr. Josef Mengele. When the Jews who had been rounded up like animals were unloaded from the cattle cars, Mengele was there to direct them left or right. The old, the infirm, and the very young went left to be eliminated, while those capable of work went to the right.

My mother ended up as a slave laborer in a Nazi factory until liberated by the Russians, but my four grandparents met their end in the gas chambers. One of my mother's sisters also was "saved" by Mengele, but only to be experimented upon. The Nazis carried out all sorts of vile experiments, often leaving the victims mutilated and in agony if they managed to survive. I don't exactly know to what horrific experiment Dr. Mengele subjected my aunt because she never talked about it, but I know that she was never able to have children.

While I had been well educated about the unspeakable horrors of the Holocaust, growing up in Montreal, I didn't encounter much dialogue about anti-Black racism. Being a baseball fan, I knew about the problems encountered by Jackie Robinson with the Brooklyn Dodgers, but I had also read that he did not face the same kind of prejudice when he played for the Montreal Royals.

Then came an eye-opening road trip through the U.S. in 1964 with a couple of friends. We had just gotten our driving licenses and one of my friends had bought a used car with his bar mitzvah money. One evening, we somehow managed to end up on a back road in Alabama and our attention was drawn to what looked like a brush fire in a field. We stopped to look. It was an astounding sight. A large cross was aflame, surrounded by a bunch of men in white cloaks and hoods. We had chanced upon a KKK rally! I can still hear the speech ringing in my ear. "You can dress 'em up, send them up North, give 'em a college education and they will still . . ." We didn't wait around to hear the rest, quickly deciding that this was not a place for us to be. I knew that the KKK existed, and I knew about slavery and the Civil War, but witnessing this organized exhibit of pure racism left an indelible mark.

My favorite subject in high school was actually history, and the Alabama experience prompted me to look more deeply into the saga of Black people in America and the rise of racism. It is a bitter story with a sweet angle. It was the appetite for sugar by white Europeans that spawned the slave trade and led to some twenty million Africans being ripped from their homeland and deposited in the Americas, where they were often treated like animals, being whipped, beaten, and branded. To the plantation owners who lived lavish lives, they were no more than property. Sickening.

The remnants of that dreadful era still haunt African Americans today, and while no longer literally enslaved, they still live with the yoke of discrimination around their neck. As we have seen, sometimes that yoke can literally squeeze out a last breath. Maybe this particular last breath will breathe life into a movement that does more than serve up pious platitudes about equality. The time has come to unleash and enforce laws against intolerance, bigotry, and overzealous police actions. The U.S. Declaration of Independence got it right with "all men are created equal." Those were just words. Forty-one of the men who signed the Declaration owned slaves. Words don't mean much unless they are acted upon.

With that being said, let's get down to science. Given how the pandemic has dominated both the scientific literature and the popular media, I do feel a need to contribute to the discourse. However, I do not want to forsake discussion of other matters, because life does go on, and there's lots of fascinating science to explore. Indeed with the scientific literature exploding, websites multiplying, and blog posts increasing exponentially, one could say that "Science Goes Viral" in more ways than one.

PART I

THE VIRUS STRIKES

DOWN THE TOILET

One of the first signs of impending doom presented by the appearance of COVID-19 was the disappearance of toilet paper from store shelves. People panicked at the possibility of their bottoms being assaulted by rough paper ripped from paper bags or magazines. Actually, there was never any need to be so spooked because even during the lockdown, paper producers were deemed to be an essential service and toilet paper was being rolled out at a normal rate. The perceived shortage did have an effect though. It focused the spotlight on this commodity as well as on the various methods to which people historically resorted in pursuit of eliminating remnants of nature's call.

In ancient times, the handiest solution was, well, the hand. Usually the left. That's why in some cultures, eating, or even just touching someone with the left hand, is still regarded as a sinister practice. The ancient Greeks used stones, while the Romans favored a xylospongium, a natural sponge on a stick inserted through a vertical opening on the front of a stone toilet seat. There was no need to stand up to finish the job. Interestingly, a modern incarnation of this device is available for overweight people, but instead of a sponge, the gadget holds a piece of toilet paper.

In Africa, the large feathery leaves of the *Peltophorum africanum* tree, also known as the "toilet tree," have a long history of use. Medieval monks used cut-up pieces of habits that had become too threadbare to wear. They referred to these as *anitergium*, from the Latin *ani* for "anus," and *tergeo*, "to scrape." In the sixteenth century, there were apparently enough rumblings about various wiping options for the Renaissance writer Rabelais to satirize them in his classic *Gargantua and Pantagruel*, a work that has a whole chapter devoted to different methodologies. His conclusion was that the best results were achieved with the head of a well-downed goose held between the legs. Rabelais claimed a further benefit in that "you will feel in your rear a most wonderful pleasure." As far as I know, no corroborating evidence is available to demonstrate the benefits of such goosing.

Henry VIII wasn't keen on taking the matter into his own hands, so he appointed a "Groom of the King's Close Stool" whose job was to unsoil the royal rear. The close stool was a throne with a hole and a chamber pot. It was equipped with a lid that could be locked to ensure that only the king's bottom would enjoy the upholstered seat. Henry's daughter, Queen Elizabeth I, followed her father's practice but changed the name of the attendant to the more delicate, "First Lady of the Bedchamber." Over in France, Louis XV's mistress, Madame de Pompadour, used only the finest French lace while Madame de Maintenon, Louis XVI's second wife, favored the ultrasoft wool of Merino sheep.

In America, the large velvety leaves of the mullein plant were popular, but the most significant American contribution was the corn cob. With the kernels removed, this proved to be so effective that even up to the 1940s it was common to find corn cobs in outhouses along with catalogues from which pages could be torn. Farmlands in Middle America were sometimes referred to as the "cob and catalogue" belt.

There are Chinese references to wiping with paper as early as the sixth century, and evidence that special tissues were produced on a large scale for the imperial court by the fourteenth century. In North America, though, toilet paper did not appear until 1857 when Joseph Gayetty introduced "paper for the water closet." Given that pages from free catalogues were readily available, he needed a marketing gimmick. The health gambit fit the bill. Gayetty claimed that Americans were ruining their physical and mental health by wiping with printed paper that had "death-dealing" chemicals such as lampblack, oxalic acid, oil of vitriol, and chloride of lime. His "Medicated Paper" was "pure as snow." Unlike catalogue paper, it wouldn't cause hemorrhoids and "would cheat physicians out of their fees." When flush toilets became more common, he cleverly changed the pitch to "this paper will dissolve so that it will not like ordinary paper choke the water pipes."

The next breakthrough came with engineer Seth Wheeler patenting the perforated toilet paper roll in 1877. Then Clarence and Irvin Scott took sales to new heights with an emphasis on softness. An ad from 1941 featured a woman's complaint to a doctor that her husband was becoming abusive. Perhaps, the doctor suggested, he may be irritated by bad bathroom tissue. The paper was too rough, which he demonstrated by showing it was opaque to light, while the light passed right through Scott's "Waldorf" paper. Switching to this "they became the happiest couple in town."

Today, in this unhappy COVID era, at least some happiness is to be found in not having to worry about running out of toilet paper. We can even have it formaldehyde-free, chlorine-free, BPA-free, unscented, and recycled. We can even squeeze the Charmin without Mr. Whipple interfering as he did in the famous ads from years ago.

CHARLIE CHAPLIN, DR. TUTTLE, AND THE SPANISH FLU

Avoid public gatherings. Close churches and theaters. Don't shake hands. Don't spit. Wear masks. Sounds like Dr. Anthony Fauci in 2020. But those words were spoken by Dr. Thomas Tuttle in 1918 when the Spanish flu, which did not originate in Spain, was sweeping across the globe. Countries involved in the First World War censored stories about the flu so as not to create even more panic than that sparked by the war. Spain was neutral, and newspapers published extensively about the flu in that country, forever linking the disease with Spain.

Dr. Tuttle was a specialist in infectious diseases and had been appointed health commissioner in the state of Washington. Although viruses had not yet been discovered, he was convinced that the disease was spread through human contact, particularly by the coughs and sneezes of people who had been infected. He even raised the prospect of individuals transmitting the disease without being sick themselves, a situation now termed asymptomatic transmission. Tuttle had learned of an outbreak of flu on a steamship traveling from Nome, Alaska, to

Seattle and wondered how that could have happened given that no cases of flu had been reported in Nome. The conclusion was that someone had boarded the ship and spread the infection despite having experienced no symptoms. Based on his observations, Dr. Tuttle advocated what today is being called social distancing and even managed to convince the mayor of Seattle to impose fines for not wearing a mask on a streetcar and for spitting on the sidewalk. Not unlike today, there was opposition to the severe measures, especially the closing of churches. The mayor responded by suggesting that "religion which won't keep for two weeks is not worth having."

With the implementation of Tuttle's recommendations, the spread of the flu in Seattle slowed significantly, and other cities took note. San Francisco imposed a five-dollar fine, heavy at the time, for "disturbing the peace" by not wearing a mask in public. St. Louis banned public gatherings and closed schools with results that sharply contrasted with cities that did not take severe measures, such as Philadelphia. The "City of Brotherly Love" went ahead with a parade to raise money for the war effort, which led to the flu spreading like wildfire. In just one month, over 11,000 Philadelphia residents died, including 759 on the worst day of the outbreak. Special wagons drove around the city with drivers hollering, "Bring out your dead!" The collected corpses were then buried in mass graves.

In New York, with the pandemic in full swing, opening night of a new Charlie Chaplin movie, *Shoulder Arms*, was for many film buffs too hot to resist, despite the city's discouraging of public gatherings. The Little Tramp kidnapping the Kaiser sounded like a welcome relief from wartime news. Many of the attendees got relief alright — they were relieved of their lives. The manager of the theater, Harold Edel, on seeing the large opening-night crowd, exclaimed, "We think it is a most wonderful appreciation of *Shoulder Arms* that people should veritably take their lives in their hands to see it." He probably did not think that one of those lives would be his. Edel died of the Spanish flu soon after mingling with the opening night crowd.

Although Seattle was successful in temporarily curbing the pandemic, Dr. Tuttle warned that once restrictions were eased, a second wave would follow, and indeed it did. He became disenchanted with how various levels of government were interested only in dealing with the problems at hand and not in preparing for the next pandemic, which he believed was inevitable. He wrote that "owing to the inclination of our government (city, county, state, and national) to provide funds for operating only when sickness is present, and to absolutely cut off any support whatsoever for the study of the epidemiology of the disease after an epidemic has passed, renders it very probable that we will meet our next epidemic (probably 20 or 30 years hence) with as little knowledge of the true nature of the disease as we had when we confronted the epidemic in the fall of 1918."

In terms of time, he wasn't off by much. In 1957, the "Asian flu" spread from China to the rest of the world, causing some 1.1 million deaths globally with about 116,000 in the U.S. But Dr. Tuttle was wrong about scientists having as little knowledge as in 1918. The electron microscope had been introduced in the 1930s, making viruses visible. By 1940, experiments were underway to produce antiviral vaccines eventually leading to one that would be successful in containing the Asian flu pandemic. There was of course no vaccine to curb the Spanish flu of 1918–19. That plague eventually abated when infected people either developed immunity or died.

And what happened to Dr. Tuttle? Even though he was a visionary in terms of the effectiveness of physical distancing, masks, and quarantine, his advocacy for essentially shutting down the city made him a very polarizing figure and resulted in him losing his job as health commissioner.

The Spanish-born philosopher George Santayana put it well in his classic assertion that "those who cannot remember the past are condemned to repeat it." So let's remember the *Shoulder Arms* disaster, and let's be very careful with rushing back to public gatherings. Luckily, technology now allows us to watch Chaplin's films without having to go to the theater. In these *Modern Times*, we can watch *The Great Dictator* in the comfort of home. Think twice about sharing the popcorn though.

PANGS ABOUT THE PANGOLIN

In 1820, Lord Francis Rawdon-Hastings, the Governor-General of Bengal from 1813 to 1823, presented King George III with a most unusual gift. On first glance, it resembled a coat of armor, but it was actually a jacket made from the scales of the pangolin. What is a pangolin? A curious-looking animal, something of a cross between an anteater and an armadillo that few people in the west had ever heard of before some researchers postulated a connection between it and the COVID-19 crisis.

There are eight species of this mammal, found only in Asia and Africa, and all are covered with scales that give the appearance of armor. When attacked by a predator, the pangolin curls up into a ball, protecting its soft underbelly. While a lion, leopard, or tiger can do no more than roll the living ball around until the predator gets frustrated and goes on to look for easier prey, the coiling up does not offer protection from the pangolin's most dangerous predator, man. The helpless animal is easily picked up and cannot even bite its nemesis since it has no teeth. It does have a very long tongue, which is fine for sucking up insects and termites but is not much of a deterrent for hunters, who club the unfortunate animal unconscious before tossing it into boiling water to loosen its scales.

Pangolin flesh is regarded as a delicacy in Asia and is thought to stimulate the sexual appetite, selling for upwards of $600 per kilo. But it is its scales that bring in the big money, fetching around $3,000 a kilo. That's because roasted and ground up, they are believed to stimulate lactation, improve blood circulation, heal menstrual problems, and cure various ailments, even cancer. None of this makes any sense given that the scales are just made of keratin, the same protein that forms the basis of our hair and nails.

Just as nonsensical is the belief that "pangolin wine," made by boiling baby pangolins in rice wine, heals skin diseases and relieves asthma. Even drinking pangolin blood is said to have medicinal value. Little wonder then that pangolins are poached to such an extent that they have become the most trafficked animal in the world! When COVID-19 first appeared in China, there was a rush for such traditional medicines, fueled by the inability of modern science to offer an effective treatment. However, pangolin products soon lost their appeal after extensive publicity was given to research that seemed to link a virus discovered in these animals to the disease.

Two Chinese scientists found that the genetic sequences of a coronavirus in pangolins matched the sequences in the SARS-CoV-2 virus to the extent of 99 percent. This raised the possibility that the virus had jumped to humans from pangolins, likely through poachers handling live animals. After the release of the preliminary results, a more detailed analysis revealed that the 99 percent match was only in one region of the virus's genome and that the overall match was only 90 percent. This seemed to absolve the pangolin from being the culprit in transmitting the coronavirus to humans and attention focused on various other wild animals that were sold in the "wet markets" in China.

The general scientific consensus is that bats are the original source of the coronavirus since a virus they harbor is a close match for the one that has infected humans. However, there are two issues here. First, humans have limited contact with bats, and second, while the bat virus genome is very similar to the genome of the virus that infects humans, it does not have the genes that code for the proteins that allow the virus to bind to human cells.

This suggests the likely existence of an intermediate host, some animal that happened to harbor a virus that had a sequence coding for the binding protein. If this virus mingled with the bat coronavirus and exchanged some genes, a novel virus capable of infecting humans could have emerged. There is actually a model for this: the original SARS virus is believed to have been passed to humans from bats through civet cats, likely through the civets being exposed to bat poop. Such a scenario is possible with the current virus as well. And pangolins now reenter the picture. While the pangolin virus as a whole is less similar to SARS-CoV-2 than the bat virus, its genome does have the crucial binding sequences. This raises the possibility that the path may have been from bat poop to pangolin to man. Research may eventually reveal if this is so or not, but the possibility of a link has already deterred people from buying pangolin products, so for this unusual animal the cloud of COVID-19 has a silver lining.

By the way, the pangolin coat presented to King George III, replete with its gilded scales, is on display in the Royal Armouries Museum in Leeds, England.

THE SCIENCE OF EPIDEMIOLOGY

Trying to solve the mystery of a disease involves asking basic questions about who is affected, when they are affected, how they are affected, and by what they may be affected. The task of finding appropriate answers is in the domain of epidemiologists — scientists who study the incidence, distribution, causes, and control of diseases.

The ancient Greek physician Hippocrates may well be regarded as the founder of this discipline, based on his attempts to describe illnesses from a rational perspective, rather than ascribing causes to supernatural effects. Human ailments were not the result of vengeful gods, he maintained, and causes and treatments were best identified through careful, systematic observation. Life-threatening fevers were more likely to afflict people who lived in swampy areas, Hippocrates observed, although he failed to make the connection to mosquitoes. That link would finally be made by American physician Dr. Walter

Reed, but not until two millennia had passed. The famous military hospital in Washington, D.C., where presidents are usually treated, is named after Reed.

Physicians should observe people's behavior, especially their dining and drinking habits, Hippocrates advised. He is therefore often credited with introducing the idea that diet can be both the cause and cure of disease. The changing seasons, winds, and vapors play a role in the spread of disease, he noted, and also suggested that the sources of water should be considered when looking at disease patterns. He was correct, but water would not be identified as a conveyor of disease until the nineteenth century. Essentially then, Hippocrates introduced the basic concept of epidemiology, namely that the key to elucidating the nature of a disease lies in the observation of all possible contributing factors.

Strangely enough, nobody donned Hippocrates's mantle until the seventeenth century when Thomas Sydenham, eventually dubbed the "English Hippocrates," published *Observationes Medicae*, in which he emphasized the importance of making observations rather than relying on ancient authorities. Curiously, one of those ancient authorities was Hippocrates, who had recommended that fevers be treated with heat — a therapy that, according to Sydenham's observations, did not work.

The English doctor noted that the rich seemed to have a higher mortality rate from smallpox than the poor and concluded that bloodletting and the various toxic potions that were the treatments for smallpox at the time, and which were not accessible to the poor, were more dangerous than helpful. He also recognized the therapeutic value of humor, writing that "the arrival of a good clown exercises a more beneficial influence upon the health of a town than of twenty asses laden with drugs." Sydenham was the first to use iron to treat iron deficiency anemia, popularized the use of quinine to treat malaria, and promoted fresh air, exercise, and a healthy diet. For pain, his solution was opium. After carefully observing its effects on his patients, he concluded that "of all the remedies it has pleased almighty God to give man to relieve his suffering, none is so universal and so efficacious as opium."

Curiously, although Sydenham promoted careful observation, he was not in favor of studying disease by means of autopsies, or even through the use of the recently introduced microscope. He was deeply religious and argued that God only gave man the ability to perceive the outer nature of things with his senses.

Epidemiology took on a decidedly scientific approach in the eighteenth century with Scottish naval surgeon James Lind tracing epidemics of scurvy among sailors to deficiencies in the diet and famously recommending the inclusion of citrus fruits as a preventative. Then in 1798, Edward Jenner made his classic observation that milkmaids previously afflicted with cowpox did not contract smallpox. This prompted him to try to protect people from smallpox by inoculating them with pus taken from cowpox pustules, marking the beginning of vaccination.

Then in the nineteenth century, epidemiology took two large leaps, as elaborated on in the following two chapters. Ignaz Semmelweis observed that women were more likely to die from childbed fever if during birth they had been assisted by a doctor rather than a midwife. After studying many cases, he noted that the increased risk was likely due to doctors unwittingly passing on some disease-causing substance picked up in the autopsy room. The solution? Handwashing!

Dr. John Snow, a contemporary of Semmelweis, investigated an outbreak of cholera in London and plotted on a map the houses where the afflicted lived. He found a cluster of the disease in and around Broad Street and traced the problem to a public water pump from which people drew water. He concluded correctly that cholera was somehow borne by the water and suggested that the handle of the pump be removed to stop the spread of the disease. This insight is widely regarded as a pivotal point in epidemiology and is celebrated every year in London with the John Snow Society's Pumphandle

Lecture, traditionally accompanied by the ceremony of removing and then replacing the pump handle as a reminder of the continuing challenges to public health. Hopefully, one day there will be a lecture on how the challenge of COVID-19 has been met.

SINGING HAPPY BIRTHDAY IS NO LONGER ONLY FOR BIRTHDAYS

Lady Macbeth is enjoying renewed popularity in light of the current coronavirus situation. "Will these hands ne'er be clean?" she asks in the famous sleepwalking scene as she mimics washing her hands. She doesn't exactly exercise the right technique, but, of course, the action is symbolic. The "damned spot" she is trying to rid herself of isn't physical, it is guilt. Today, handwashing has a different kind of guilt associated with it. That guilt descends if we don't sing at least two stanzas of "Happy Birthday" as we lather with soap, scrub the backs and palms, twist a thumb as we grip it with the other hand, or ensure that our nails have received enough attention.

Although soap was well-known by 1606 when Shakespeare wrote his classic play, it was not commonly used. Sanitation was not a component of life. Any knowledge that invisible microbes could transmit disease would not emerge until the nineteenth century. However, even before Louis Pasteur laid the foundations for the germ theory of disease in the 1860s, a Hungarian physician hit upon the importance of washing hands to prevent disease transmission.

As a young doctor, Ignaz Semmelweis was keenly aware of childbed fever. It was not unusual for a mother to die within a week of giving birth, but he did note that more women were dying after giving birth if they were attended by doctors than by nurses. Semmelweis became obsessed with this conundrum. He performed numerous autopsies on the dead women in a search for some causative agent but found none. Then in 1847 came a tragic breakthrough. One of his colleagues cut himself during an autopsy and soon died of symptoms that were remarkably similar to childbed fever. Semmelweis surmised that some sort of cadaver particles must have gotten into his friend's bloodstream and killed him. And perhaps these same cadaver particles were also killing the women! Now the difference between the two obstetrics wards became clear. The doctors who assisted in the births in the infamous death ward, and who performed internal exams on the women before and after birth, often came directly from the autopsy room where they were trying to solve the horrific problem of childbed fever. Could they be infecting their patients with some sort of cadaver particles? This now seemed possible. After all, doctors' hands constantly smelled of cadavers.

The conclusion now seemed obvious. Semmelweis urged all doctors and students to thoroughly wash their hands after performing autopsies. But even with thorough washing, a faint smell of the autopsy room persisted, so he decreed that the hands should be rinsed in a hypochlorite solution. Hypochlorite bleach at the time was already known to eliminate smells, although why it did so was not understood. The results of the handwashing bordered on the miraculous. Within a year, the death rate fell from a high of 30 percent to 3 percent. The notorious "death ward" was no more. Semmelweis was elated by this result, but he was also troubled by it. He realized that he himself had probably been responsible for many deaths as he rushed back and forth between the obstetrics ward and the autopsy room. His feelings of guilt, coupled with his conviction that he had made a major discovery, converted Semmelweis into a hand-wash-promoting zealot. Still, it took decades before the importance of handwashing took hold in the medical community.

Today, with our extensive knowledge of disease transmission by microbes, it is clear why handwashing works. Bacteria and viruses are either inactivated or rinsed away. Soap molecules have one end that is soluble in water and another that dissolves in fatty substances, or "lipids." Most dirt is of an oily or greasy nature and attracts the fat-soluble end, leaving the other end to be anchored in water. Rinsing then pulls the oily dirt off any surface to which it is attached. In the case of COVID-19, soap can actually destroy the virus responsible for the disease. Coronaviruses are composed of a core of nucleic acids, either RNA or DNA, surrounded by a protective coating made of proteins and fats. The fat-soluble end of the soap molecule embeds itself in the lipid layer and the virus is then literally pulled apart, since the rinsing water is tugging the other end. The reason for the 20–30 second time period is to ensure that the soap makes contact with whatever microbes may be present.

Washing with soap and water is more effective than using an alcohol-based hand sanitizer. Alcohol can dissolve fats, so it is capable of stripping away the lipid layer of a virus, thereby inactivating it, but the problem is that unlike washing with soap it doesn't remove dirt and may not get at viruses that are stuck in the dirt. Of course, when soap and water are not available, hand sanitizers can step in as long as they contain at least 60 percent alcohol.

Being urged to stay home because of this damned virus, why not make use of the opportunity and find a version of Macbeth to watch? I don't think "eye of newt and toe of frog, wool of bat and tongue of dog" is the solution to COVID-19, although equally nonsensical regimens are being peddled by the charlatans who emerge out of the woodwork whenever a crisis such as this presents.

OH MY, MIASMA

In 430 BC, a great plague struck the city-state of Athens, killing about a quarter of the population. Historians have proposed smallpox, bubonic plague, typhus, and even Ebola as possible causes for the devastation, but there is no conclusive evidence for any of these. As far as the Athenians were concerned, they believed that the gods had turned against them. Hippocrates, the "father of medicine," did not concur. His view was that all forms of illness had a natural cause, such as foul air emanating from putrefying carcasses or rotting vegetation. Epidemics, he suggested, could be curbed by lighting fires of aromatic wood in the streets. An interesting idea, but futile, although the pleasant smell may have at least masked some of the putrid odors wafting out from the funeral pyres burning all over the city.

Hippocrates's ideas about noxious vapors causing disease would eventually become the "miasma" theory of disease, the term deriving from the ancient Greek word for defilement. Malaria, for example, from the Latin *mala* for "bad" and *aria* for "air," was thought to be caused by inhaling foul vapors. The infection does sort of come from the air, although the culprit is a parasite borne by mosquitoes, not some ill wind. Amazingly, Hippocrates was so revered that the miasma theory held sway until Louis Pasteur managed to link microbes with illness and spawned the germ theory of disease.

Actually, while Pasteur did prove that germs can cause disease, he wasn't the first to propose that sickness can be spread by some sort of invisible agent. In the eleventh century, the Persian physician Avicenna suggested that people can spread disease to others through their exhaled breath; he even discussed transmission by water and dirt. Then in the fourteenth century, Arab physicians Ibn Khatima and Ibn al-Khatib hypothesized about "minute bodies" passing disease via clothing, jewelry, or shared food. Two hundred years later, Italian physician Girolamo Fracastoro proposed that diseases are spread by "seed-like particles" that can transmit infection by direct contact with a sick person, or even without contact over long distances.

These minute bodies and seed-like particles became a reality in the seventeenth century when Antonie van Leeuwenhoek peered through a microscope of his own design and observed little "animalcules" cavorting around. He didn't link these to disease, but the germs, as they came to be called, played a significant role in Dr. John Snow's classic 1849 essay "On the Mode of Communication of Cholera." Although Pasteur would not publish his germ theory until 1861, it is clear that Snow had an understanding of disease transmission by tiny invisible entities. He correctly suggested that cholera was transmitted by the fecal-oral route and hypothesized that "the excretions of the sick at once suggest themselves as containing some material which being accidentally swallowed might attach itself to the mucous membrane of the small intestines, and there multiply itself by appropriation of surrounding matter."

When cholera broke out in London in 1854, Snow proved his point by demonstrating that people who obtained their water from a company that sourced it from within the city had a death rate of 315 per 10,000 households, while those who purchased water drawn from a cleaner part of the river died at a rate of just 37 per 10,000. Then came his most famous contribution, mapping the incidence of cholera in Soho and revealing a cluster of cases around the water pump on Broad Street. Despite the evidence, a parliamentary committee charged with examining the cholera epidemic discounted Snow's germ theory in favor of the miasma theory, claiming that effluvia from "offensive trades" such as bone boilers and tallow producers were the actual cause of the disease.

The prestigious medical journal *The Lancet* published a scathing editorial insinuating that Snow was in the pocket of the industries that produced "pestilent vapors, miasms, and loathsome abominations of every kind" and described his conduct as "perverse, crotchety, or treasonable." Although the parliamentary committee was wrong about the transmission of cholera, it did do some good by recommending that industries producing noxious fumes be regulated and that proper methods of disposing of sewage be instituted.

Dr. Snow unfortunately did not live to see the day in 1884 when microbiologist Robert Koch isolated the bacterium causing cholera, proving that his theory of transmission by a germ had been correct. To be fair, the miasma theory was not totally wrong. The World Health Organization estimates that currently some 4 million people die every year due to air pollution. The main culprits are nitrogen and sulfur oxides from combustion of fossil fuels, ozone from the effect of ultraviolet light on nitrogen dioxide, and hydrocarbons, passive smoking, and the burning of biomass. I suppose that aerosols carrying the SARS-CoV-2 virus may also fall into the category of "defiled air," so in a sense COVID-19 conforms somewhat to the miasma theory.

BOOSTING IMMUNITY

"Reinforce and Boost Your Immune System," promised the ad splashed across two pages of a local newspaper. There were products galore! Mushroom extracts, probiotics, collagen supplements, exotic oils, bee propolis, and various "kefir-kombucha fermented blends." No direct reference was made to COVID-19, but the implicit message was clear. If somehow we can boost our immune system, we will be in a better position to ward off this nasty virus. Sounds good, but the truth is that "boosting the immune system" is a scientifically meaningless claim. So is "supports immune function." These are marketing terms, not scientific ones.

The immune system is not like a muscle that we can build up through exercise. It is a highly complex system of cells, tissues, organs, and a host of chemicals they produce to help the body fight infections and other diseases. There are two basic components to the immune system, innate and acquired; the latter is sometimes called adaptive. The innate system swings into action with a variety of cells and chemicals as soon as the body is attacked by a foreign substance, be it a bacterium, a virus, or toxin. The reason we sneeze and cough and have a runny nose when we catch a cold is that the body is trying to expel the intruding virus. Our throat may become inflamed because blood vessels dilate, allowing white blood cells to rush to the scene to engulf and neutralize the invader. That's our innate immune system in action.

But thanks to the acquired or adaptive immune system, we never catch the same cold twice. That's because exposure to an intruder stimulates the immune system to produce antibodies, special proteins that recognize the intruder on a subsequent exposure and neutralize it. This is an extremely oversimplified view because there are dozens of different white blood cells, numerous antibodies, and all sorts of messenger chemicals involved in the proper functioning of the immune system. It is true that a poor diet, lack of exercise, impaired sleep, stress, and aging can diminish the immune response, but this cannot be remedied with supplements, juices, chiropractic adjustments, or homeopathic treatments.

Even if immunity could be boosted in some simple fashion, such an intervention could have negative consequences. An overactive immune system can attack healthy tissues, which is exactly what happens in autoimmune diseases such as arthritis and lupus. There is increasing evidence that conditions such as diabetes and heart disease are associated with low-grade chronic inflammation that is a result of unwelcome immune activity. Basically, we want our immune system to protect us from foreign substances, but we do not want it to go into overdrive and attack healthy tissues. Luckily, when in a healthy state, our bodies manage this well.

In fact, one of the most fearsome complications of COVID-19 is a cytokine storm, which is a hyperimmune reaction. Cytokines are molecules that are released when the body senses an invader, but if they become too abundant, the immune system may not be able to stop itself. Immune cells spread beyond infected body parts and start attacking healthy tissues, gobbling up red and white blood cells and damaging the liver. Blood vessel walls open up to let immune cells into surrounding tissues, but the vessels get so leaky that the lungs may fill with fluid, and blood pressure drops. Blood clots throughout the body, further choking blood flow. When organs don't get enough blood, a person can go into shock, risking permanent organ damage or death.

The question to ask when a claim is made about boosting the immune system is, What exactly is being boosted? White blood cells? Which type? Macrophages? Eosinophils? Killer T cells? Helper T cells? B cells? Lymphocytes? Chemical messengers? Which ones? Interferons? Interleukins? Cytokines? Tumor necrosis factors? How about antibodies? IgG? IgA? IgM, IgE, or IgM? Where are the studies that show that taking any supplement, be it kombucha or vitamin C or green tea, has any effect on any of these parameters? There are none! The exception may be vitamin D, with low levels being linked with a higher risk of respiratory infections and the development of autoimmune diseases.

Oh yes . . . there is one way that we can boost immunity. Vaccination! That actually can be shown to produce antibodies against invading organisms.

FROM VARIOLATION TO VACCINATION

The noose was waiting for the six criminals being held in Newgate Prison in 1722. Understandably, they jumped at the chance of being pardoned and released if they would take part in an experiment. Through a scratch on their arm, they would be inoculated with pus from the scabs of a smallpox victim and would later be exposed to the disease to see if they had developed any immunity.

Dr. Charles Maitland had been granted a Royal License to perform the experiment at the urging of Caroline of Ansbach, then Princess of Wales. A year earlier, Caroline, along with three physicians, had been impressed as Maitland performed the procedure on the four-year-old daughter of Lady Mary Montagu, who had recently returned from Turkey after spending several years there as the wife of the British ambassador.

In Constantinople, Mary had learned about the ancient practice of ingrafting to prevent smallpox. She famously described this in a letter to a friend in 1717. "The old woman comes with a nut-shell full of the matter of the best sort of smallpox and asks what veins you please to have opened, and with a large needle puts into the vein as much venom as can lie upon the head of her needle; there is no example of anyone that has died."

Mary, who had been left with a severe case of facial scarring after having survived smallpox herself, was so taken by the possibility of preventing the disease that she had her son inoculated under the supervision of Dr. Maitland, who then was serving as physician at the British Embassy. The boy had no significant aftereffects, and upon their return to England, both Mary and Maitland became enthusiastic supporters of ingrafting. It was their advocacy that led to the Newgate experiment, which all the prisoners survived. Princess Caroline then suggested that six orphan children be given the same treatment as a further test, and the results were so satisfactory that Caroline had her children inoculated against smallpox. The royal family went on to promote the procedure throughout England.

In America, the practice of infecting a person with an exudate from a patient with a mild form of smallpox was introduced by Dr. Zabdiel Boylston, prompted by an outbreak of smallpox in Boston in 1721. He was convinced to try the procedure by Puritan minister Cotton Mather, who in turn had learned about it from his slave, Onesimus, who had been inoculated in Africa. Boylston's approach was not well received, with some opponents invoking Jesus's statement in the Bible that "it is not the healthy who need a doctor, but the sick." There were even threats on the doctor's life, forcing him to visit his patients in disguise. But Boylston and Mather confronted the hostility with science. They collected information about the number of smallpox cases and showed that inoculation dropped the death rate significantly. After this, resistance faded but did not disappear.

The idea of ingrafting traces back to the tenth century in China, where physicians introduced the practice of blowing dust made from the dried scabs of smallpox victims up the nose of healthy people. This was prompted by the observation that an individual who had survived smallpox did not contract the disease again. The process, and the ingrafting later introduced by the Turks, triggered, at least in most cases, a mild form of smallpox that resulted in immunity against the disease.

Today, we know that smallpox is a viral disease and that exposure to a limited amount of virus prompts the body to form antibodies that destroy the virus on any subsequent exposure. Ingrafting was not risk-free, since some people did develop a serious case of smallpox, and even those who had only mild symptoms were still able to infect others even though they went on to develop immunity. The treatment eventually faded when Edward Jenner, who himself as a boy had been subjected to ingrafting, discovered that inoculation with extracts of pustules from people who had contracted cowpox, a milder viral disease, offered safer and more effective protection against smallpox. Jenner's "vaccination," from the Latin for "cow," met with opposition, but eventually the smallpox vaccine proved its mettle by eradicating the disease from the world.

Once it was discovered that smallpox was a viral disease — the virus responsible is named *variola* from the Latin for "pox" or "pustule" — the process that had been referred to as inoculation or ingrafting was retroactively renamed "variolation." Today, this term is being bandied about in conjunction with COVID-19, with some researchers proposing that the wearing of face masks is a form of variolation that may lead to a less severe form of the disease.

The argument is that a mask reduces the number of viral particles that may be inhaled to the extent that any disease triggered would be mild and likely asymptomatic. The case is bolstered with reference to a cruise ship where everyone was masked, and although there were some infections, 81 percent of these were asymptomatic, as compared with 20 percent in cruise ship outbreaks without universal masking. While variolation via masking may be a tenuous contention, if it helps convince some people about the benefits of wearing masks, it is welcome. Of course, as we have seen for ingrafting and vaccination, masking meets opposition in spite of all the evidence that science can muster.

A VACCINATION TRIUMPH

Movie theaters and swimming pools were closed, public gatherings were canceled, the sick were quarantined, special hospital wards were opened to care for the afflicted, sales of disinfectants soared, and police officers intercepted travelers. COVID-19? No. The flu pandemic of 1919? No. This was 1916 and New York was in the midst of an epidemic of polio, then known as "infantile paralysis."

A decade earlier, Austrian scientists Karl Landsteiner and Erwin Popper had determined that the disease was caused by a virus, but exactly how it was spread was not known. New Yorkers accused cornflakes, electrical disturbances, mercury, and cookware of being possible culprits. Even cats were suspected of being disease vectors. When a rumor emerged that authorities had placed a bounty on cats, some 70,000 of the animals were rounded up and dispatched by boys who hoped to cash in. Treatments with oxygen, quinine, radium water, mustard poultice, and even chloride of gold were tried without success. As winter approached, the epidemic subsided, only to emerge again in subsequent years, with particular venom in the 1950s.

The public was terrified. Pictures of the ill confined to iron lungs, which took over the work of paralyzed respiratory muscles, dominated newspapers. Franklin Delano Roosevelt's polio diagnosis in 1921 had shown that everyone, irrespective of social status, was susceptible to this incurable disease. A survey in 1952 found that Americans feared only nuclear war more than polio! Since infections surged in the summer, as did flies and mosquitoes, an association seemed possible. This led to widespread spraying of DDT in hopes of banishing the disease. It didn't.

Then came a pivotal moment. On April 12, 1955, ten years to the day after the death of President Roosevelt, reporters were invited to a press conference at the University of Michigan. Some 54,000 physicians around the country also had their eyes glued to a closed-circuit telecast as Dr. Thomas Francis Jr., director of the Poliomyelitis Vaccine

Evaluation Center, strode to the podium. He began by announcing that "during the spring of 1954, an extensive field trial of the effectiveness of a formalin-inactivated poliomyelitis vaccine, as developed by Dr. Jonas Salk and his associates, was initiated by the National Foundation for Infantile Paralysis." Then, after a dramatic pause, he continued: "The vaccine works. It is safe, effective, and potent!"

Amid an explosion of flashbulbs, and wild cheers and applause, Dr. Salk was then invited to the stage to describe the double-blind, placebo-controlled trial that had involved close to two million children. Nobody could doubt the results. To allay fears, Salk described how he and his wife and children had been the first humans to be inoculated after the vaccine had been tested on thousands of monkeys. Within a year, the Salk vaccine was approved, and within five years, polio was on the verge of being eliminated in North America.

The testing and later the distribution of the vaccine had not been totally without problems. In 1955, ten children died and some 200 were left with varying degrees of paralysis after being inoculated with a virus, prepared by Cutter Laboratories in California, that had not been properly inactivated. The polio epidemic hit Canada hard and there was a vaccine shortage, not unlike the COVID-19 situation. Montrealers were desperately waiting for a shipment in 1959 from Connaught Labs in Toronto, the major manufacturer. The supply finally arrived, but before it could be distributed, thieves broke into the storage facility at the University of Montreal and made off with 75,000 vials after locking a guard in a cage full of monkeys. After finding that the vaccines could not be easily sold, one of the thieves anonymously tipped off the police about their location. The vials were recovered and Montreal's children were inoculated, putting an end to the last great polio epidemic in Canada.

Dr. Salk's research had been inspired by the 1918–19 influenza pandemic and resulted in his developing the first effective flu vaccine, under the mentorship of Dr. Francis. The virus that caused the disease was inactivated with formaldehyde but was still recognized by the immune system as an invader, triggering the formation of protective antibodies. Salk believed the same strategy would work for polio. The virus he used had been cultured in monkey kidney cells, a technique based upon Thomas Weller, John Enders and Frederick Robbins's 1949 demonstration that the poliovirus could be grown in the laboratory using skin and muscle tissue from human embryos. Before this, the virus had to be grown in live monkeys.

Not everyone agreed that a "killed" virus was the way to go. Other researchers, Dr. Albert Sabin for one, were of the opinion that a "live" virus that had been weakened, or attenuated, would confer longer-lasting immunity. Such an attenuated, less virulent strain could be produced by passing the virus repeatedly through rodents. Sabin had also discovered that the polio virus first infected the small intestine, where it multiplied, and conjectured that an attenuated virus administered orally would also multiply there the same way and provoke an antibody response.

Like Salk, Sabin showed confidence in his vaccine's safety by swallowing some of the attenuated virus and then paid thirty prisoners to do the same. All developed antibodies and none became ill. However, he had a difficult time trying to organize large trials in the U.S. because the subject pool had been reduced by Salk's extensive experiments, and funding was also difficult to obtain. Surprisingly, the Soviets stepped in to help. They had been coping unsuccessfully with polio and offered to try Sabin's oral vaccine. By 1959, some ten million Soviet children had been successfully vaccinated, leading to Dr. Sabin's work being recognized with the Soviet Union's highest civilian honor, the medal of the Order of Friendship Among Peoples.

The Sabin oral polio vaccine is now the most widely used one in the world, largely because of the ease of administration. This despite a small risk that the virus can revert to an active form in the digestive tract and cause disease. On the other hand, after Sabin's oral polio vaccine, people shed some of the weakened virus in their feces. This exposes others in the community to the virus and boosts immunity. Canada and the U.S. judge that the risk of contracting polio from the Sabin vaccine, albeit very small, is not worth taking and use the Salk vaccine.

Whether it is Salk or Sabin who should get credit for conquering polio has been much debated, even between the two pioneers. They had a rather acrimonious relationship, each one claiming the superiority of their own approach. Certainly, the triumph over polio, one would think, is worthy of a Nobel Prize, but neither Salk nor Sabin was recognized.

Documents in the Nobel archives suggest that the committee believed that while the contributions had great practical significance, they were not "primary" but depended on accomplishments of others such as Weller, Enders, and Robbins, who received the 1954 Nobel prize in Medicine and Physiology for "their discovery of the ability of poliovirus to grow in cultures of various types of tissue and it was this milestone that led to the production of both 'killed' and 'live' poliovirus vaccines allowing Salk and Sabin to make their discoveries." Although there are questions about the allotment of credit, there is no question that the development of vaccines against polio represents one of the great triumphs of science.

COVID VACCINES, HERE WE COME!

As soon as Chinese researchers published the genomic sequence of the virus that was causing a novel type of pneumonia in Wuhan in January 2020, the race for a vaccine began. Given that no vaccine had ever been produced in less than four years, the challenge was a mighty one.

In theory, designing a vaccine seems quite straightforward, but theory is not the same as practice. The basic idea is to trick the body into mounting an immune reaction against the virus without exposing it to "live" virus. An immune reaction has essentially two components. White blood cells called B cells produce antibodies, which are specific proteins capable of binding to and neutralizing the virus. Another type of white blood cell, the T cell, seeks out and destroys cells that have been infected by a virus. This prevents the virus from replicating inside the cell, spilling out, and infecting other cells.

There are four different technologies that are generally considered by vaccine researchers. The simplest idea is to introduce the virus itself into the body, let the body recognize it as a foreign invader, and start producing antibodies and T cells that then stay in the system, ready to swing into action in case of a future attack by the same virus. Of course, this priming of the system has to be done without causing disease through the introduction of a virus capable of reproducing. This requires either inactivating the virus by destroying its genetic machinery with chemicals such as formaldehyde, or attenuating it by introducing it into a foreign host such as an embryonated egg or a live animal. In this case, the virus adapts to the host and loses its ability to infect humans. Whether killed or attenuated, the virus retains its surface proteins, which are the antigens that are recognized by the immune system.

For coronaviruses, these surface proteins take the shape of spikes that the virus uses to attach to receptors on the cells they eventually infect. These receptors are also proteins, known as ACE2 (angioconverting enzyme 2), and attachment to them is a necessary prelude to the virus invading a cell and hijacking its reproductive machinery to produce more viruses. Since the spike protein is what the immune system recognizes as the foreign invader, another possibility for a vaccine is the introduction into the body of just the spike protein detached from the virus. Since the entire genetic code of the SARS-CoV-2 virus is known, researchers were able to identify the genes that code for the spike protein. Through genetic engineering, these genes can be inserted into the genome of cells, which can then be grown in the lab to crank out a protein that can be used for producing what is called a subunit protein vaccine.

A third method, discussed in detail in the next section, aims at activating the immune system through exposure to spike protein that is actually produced in the body by cells that have been equipped with the genetic instruction to crank out this protein. The necessary genes are introduced either into DNA or messenger RNA, nucleic acids that can then be delivered by vaccines and taken up by cells. This is the scheme used both by the Moderna and Pfizer vaccines, which were first to the market based on encouraging trials that showed over 90 percent efficacy in preventing infection.

Yet another technique uses viral vectors to introduce the genes needed to produce the spike protein. A harmless virus, such as a chimpanzee adenovirus that cannot replicate in humans, has the gene that codes for the spike protein implanted in its genome. When this virus infects human cells, it transfers the gene into the cell's DNA with the result that the cell starts to churn out the protein that triggers the formation of antibodies and production of T cells. This is the approach taken by AstraZeneca in cooperation with Oxford University. The Russian Sputnik V vaccine is also based on a viral vector and uses two different adenoviruses for the required two doses. This reduces the risk of having an adverse reaction caused by sensitization to the inactive components of the first dose of the modified adenovirus.

There is no question that remarkable progress has been made in developing a number of vaccines against COVID-19 in a very short time. While production methods can be sped up, long-term effects, both in terms of efficacy and side effects, cannot be known until a long term has passed. Proper testing for both safety and efficacy takes time. Years, not months. The path to a useful vaccine is lined with curves, pitfalls, and possible dead ends.

HERE IS THE MESSAGE ABOUT MESSENGER RNA

No, we do not become genetic mutants when we roll up our sleeves and agree to be jabbed with a vaccine containing modified messenger RNA (mod-mRNA). That scare made the rounds as soon as the vaccine rollout began. The term "modified" triggers alarm bells for some people, especially those who have bought into the often-misleading information about genetically modified organisms (GMOs) spread by activist groups. Fearful that their genes may somehow be modified by the vaccine, they conjured up memories of the fate that befell the Teenage Mutant Ninja Turtles. I think we can dismiss that fear, but as with any novel pharmaceutical intervention, there are legitimate concerns. First, some background.

Just about everything that happens in our body relies on the family of molecules we call proteins. Indeed, the term "protein" derives from the Greek for "first place." Proteins are polymers, long chain-like molecules with the component links being amino acids. Since the chains can be of various lengths, with all sorts of twists and turns, and feature twenty amino acids that can be incorporated in diverse sequences, the number of possible protein structures is immense. Some proteins are the building blocks of muscles, some make up our hair and skin, some play key roles in immune function, some are hormones, and others act as enzymes, the biochemical catalysts that control the myriad chemical reactions that go on in our body and together constitute life. The instructions for making all these proteins are encoded in molecules of deoxyribonucleic acid (DNA) found in the nucleus of every cell. DNA is indeed the blueprint of life.

DNA does not synthesize proteins directly, rather it transfers the instructions for assembling the amino acids to molecules of messenger ribonucleic acid (mRNA). These molecules then migrate out of the cell's nucleus into the cytoplasm where they serve as a template for constructing proteins by little factories called ribosomes. The requirement for protein synthesis can now be seen as the presence of the appropriate mRNA in the cytoplasm.

What if that mRNA could be synthesized outside the body, in a lab, and be delivered directly into the cell? The protein that RNA codes for would be produced just as if the mRNA had come from the cell's nucleus bearing instructions from DNA. If that synthetic mRNA codes for the production of the spike protein found on the surface of the SARS-CoV-2 virus, then this protein will be produced and will go on to stimulate the production of antibodies against it by B cells of the immune system. Should the coronavirus then at some future time invade the body, the antibodies will recognize its spike proteins and latch onto them, preventing them from interacting with receptors on the surface of cells. Since that interaction is a necessary prelude for the virus's entry into a cell, infection is avoided. This is the principle behind both the Pfizer and Moderna vaccines.

The name "Moderna" derives from "modified RNA," which brings us back to the question of what modified means in this context. RNA is a polymer, the building blocks of which are simple molecules called nucleosides, analogous to how proteins are composed of amino acids. It is the sequence in which the four possible nucleosides — namely adenosine, guanosine, uridine, and cytidine — are strung together through phosphate linkages that constitutes the code for protein synthesis by ribosomes.

When Chinese researchers published the entire genome of the SARS-CoV-2 virus, attention was immediately drawn to the sequence of nucleotides that code for the spike protein. This was regarded as the best candidate as the target for a potential vaccine since it was a good bet that the virus uses this protein to gain entry into cells where it could then hijack the cell's reproductive machinery and begin to replicate. Now the task was to synthesize the mRNA that codes for the spike protein and introduce it into the body, essentially tricking the immune system into believing it has been attacked by a virus, resulting in antibody formation.

These days, building RNA from nucleotides in the lab is not a major problem. What is needed is a DNA template that can be derived from the decoding of the genome of an organism and some RNA polymerase enzymes, the same enzymes cells themselves use to synthesize RNA. There is no doubt that science has come a long way since Dr. Kelvin Ogilvie at McGill and his associates at BioLogicals developed a "gene machine" in the 1980s capable of joining a few nucleotides together using automated chemical techniques. That first historic machine now sits in a display case just outside my office, and I get to admire it every day!

Synthesis of the required mRNA was only the first step. The challenge was to deliver this relatively unstable molecule into cells intact without causing any adverse reaction, given that any foreign substance, including synthetic mRNA, that intrudes into the body can trigger an attack by the immune system. Indeed, mice reacted adversely when in initial experiments they were treated with synthetic mRNA. Katalin Kariko, a Hungarian biochemist who came to the U.S. in 1985 and eventually took up a position at the University of Pennsylvania, was intrigued by this immune reaction. She wondered whether it was a specific part of the RNA molecule that was recognized by immune cells as the enemy.

It turned out that one of the nucleosides, uridine, was the problem. Could simply replacing this with the chemically similar pseudouridine, which also occurs in the body, be the solution? Working with colleague Drew Weissman, Kariko found that this alteration did the trick. The ability of this modified mRNA to synthesize proteins was actually enhanced and the immune reaction to it was significantly reduced. Furthermore, the modified RNA was more resistant to enzymes that normally degrade mRNA. Modifying RNA in this fashion was a major breakthrough and led to other similar modifications that were critical to the development of the Pfizer and Moderna vaccines.

While RNA can exit from a cell's nucleus, it cannot enter it, and even if it could, there is no way for our genes to be "modified," since mRNA does not incorporate into DNA. Aside from RNA's instability and possible activation of unwelcome immune reactions, there was another problem. RNA does not easily penetrate the cell membrane, which is made of fatty substances. That problem was addressed by Dr. Pieter Cullis of the University of British Columbia and MIT biomechanical engineer Robert Langer, who was a co-founder of Moderna. They independently found that encapsulating RNA in various fatty substances would allow passage into the cell's cytoplasm. This paved the way for the lipid nanoparticle encapsulated mRNA used in the novel COVID-19 vaccines. The components of the nanoparticles, as well as the mRNA, are sensitive to heat, which is why these vaccines have to be maintained at a very low temperature.

While the concern about the vaccine converting us into genetic mutants can be dismissed, there is legitimate unease about some other issues. There have been a few cases of immediate, severe adverse reactions, mostly in subjects who had a history of allergic reactions. The question is, What are they reacting to? It is extremely unlikely to be the spike protein, because it takes time for the mRNA to enter cells and begin to produce it. The lipid nanoparticle contains four components, and a finger has been pointed at polyethylene glycol (PEG) as a possible allergen. However, this is used in many pharmaceuticals and cosmetics, including in large doses in laxatives. Reaction to glycols is known, but rare. That leaves the possibility of immunological reactions against the active ingredient, the mRNA itself. Although, as mentioned, nucleoside modification reduces this possibility, it does not eliminate it.

As far as possible long-term consequences go, there is just no way to know for sure before that long term has passed. Going by history, the most serious reactions to a new vaccine occur within weeks, not years after being administered. Can the spike proteins being produced present a problem down the road? This is unlikely since the mRNA that produces them breaks down quite quickly, meaning that there is no long-term production of these proteins.

Finally, all these questions have been addressed by experts in the field, both within and outside the pharmaceutical companies, with the consensus being that the benefits outweigh the risks. Admittedly, with complex situations such as the efficacy and safety of vaccines, there has to be some educated guesswork based on short-term trials and existing knowledge about immunology, physiology, molecular biology, and pharmaceutical chemistry. In any case, the brilliant scientific wizardry involved in the remarkably speedy development of the current vaccines is admirable. There are even whispers of a Nobel Prize in the offing.

MUSING ABOUT MUTATION

One summer back in high school, and yes, I can still remember that far back, I had a job running knitting machines, mostly producing furniture fabric. It wasn't particularly interesting or difficult. The machines pretty well ran themselves with my major task being to replace the different-colored spools of yarn when they ran out. Sometimes, though, the threads got tangled and the fabric came out rumpled. I'd have to stop the machine, untangle the yarn, clean the feeder, the yarn guide, and the rollers before starting up again. One day, I made a mistake and replaced a spool of polyester with a spool of Lycra yarn. As a result, the pattern on the fabric now had a sheen to it that I thought for sure would be noticed, and I'd get into trouble. It was noticed alright, but the boss liked it! And from then on we started to use some Lycra in the fabric.

Why am I recalling this now? No, it isn't really a cherished memory. I thought of it because I was trying to come up with an appropriate analogy to explain mutation, a term that has been plastered all over the media in light of the discovery of a variant of the SARS-CoV-2 virus that first surfaced in the U.K. The word mutation derives from the Latin "mutare," meaning "to change." My erroneous replacement of the spool could therefore be regarded as a mutation because it led to a change in the fabric being knitted by the machine. In this case, the mutation turned out to be beneficial since an improved product was produced as a result of the change. But there were other instances when the change was not desirable. Sometimes the belt on one of the pulleys in the machine slackened, resulting in an irregular weave and discarding of the fabric. So, mutations can be for the better or worse.

Now let's get down to the matter at hand. What happens when a virus undergoes a mutation? The SARS-CoV-2 virus is an RNA virus, meaning that all its functions are governed by the exact composition of a long, coiled molecule called ribonucleic acid found inside the viral particle. RNA serves as the blueprint for the virus's production of the proteins that are critical to its structure, as well as for those it needs to hijack a host cell's genetic machinery, allowing the virus to replicate.

An apt analogy for RNA is a string of beads, each of which can be one of four different colors. In the case of SARS-CoV-2, there are some 29,000 such beads strung together, with the beads being units called nucleotides. At various positions along the chain, there are explicit sequences of nucleotides that encode the instructions for which amino acids are to be put together, and in what sequence to form a particular protein. The identity of each amino acid that is to be included in the protein is determined by a specific set of three RNA beads in a row. The spike protein of the coronavirus, which is composed of roughly 1,300 amino acids, would therefore be encoded in an RNA sequence composed of 3,900 nucleotides. This is the protein the virus uses to latch onto receptors on a host cell, a necessary prelude to invading that cell. It is also the target of vaccines that aim to stimulate the production of antibodies that clasp the spike protein and prevent it from docking with a receptor, thereby preventing its entry into a cell.

A change in the structure of that spike protein can conceivably alter the way it engages with a receptor, in much the same way a change in a key can alter the way it fits into a lock. Of course, it depends on where the change occurs. If there is a nick in the head of a key, its function is not affected. However, even a small change in the "cuts" along the shaft can prevent the key from working. When we talk about a mutation in a virus, we refer to some change in its nucleotide sequence that will then cause a change in a protein. How does such a change come about?

Once a virus has successfully invaded a cell, it begins to replicate. That is a complex process, and errors can occur as RNA is replicated. Just a small alteration in nucleotide sequence can result in one amino acid in a protein being replaced by another. This may or may not be of any consequence, depending on where in the chain of amino acids the change occurs.

As early as June 2020, researchers detected a new variant of the virus in which one single amino acid, aspartic acid, was replaced by glycine in the spike protein. This did not seem to be a highly significant mutation because the change occurred in a part of the spike protein that was not involved in fitting into a receptor. Since then, many mutations have been noted, which is not at all unexpected. However, the U.K. variant may be a different matter because it involves the replacement of asparagine by tyrosine in a "key" part of the spike protein. This appears to result in a better fit with receptors on the host cell, making the virus more infectious. That, though, does not necessarily mean it causes more serious disease. It is also possible that the virus is more infectious because it replicates more quickly, or that it is shed for a longer time from infected people.

Whether this novel variant will be more resistant to vaccines isn't yet determined, but that is unlikely since the vaccine produces a number of antibodies that bind to different parts of the spike protein. Obviously, new variants will emerge in the future because with every replication of the virus, there is a chance of a mutation and some of these will become established in the population, as has already happened with the U.K. variant. Fortunately, most of the mutations do not alter the virus's effect on people.

And with that, I hope I have woven the fabric of a story for you that is understandable. There is no doubt, though, that some threads will surely snag and the fabric will mutate as it continues to be woven. Stay tuned.

SHARKS AND VACCINES

It was once characterized as the "Strangling Angel of Children." Diphtheria is a bacterial infection transmitted via respiratory secretions spread through the air. The bacteria produce a toxin that causes a thick film to develop in the throat, making breathing difficult, and in the worst case, strangling the patient. The name of the disease derives from the Greek for "leather," an apt description of the distinctive coating seen in the victim's throat.

In 1878, Britain was shocked by the diphtheria deaths of Alice, Queen Victoria's thirty-five-year-old daughter, and Alice's youngest child. Apparently the disease could be spread by the innocent kiss between a mother and child, the "kiss of death." The tragic case invigorated research, and within a few years, the bacterium responsible for diphtheria was isolated, and the toxin it produced identified as a specific protein. Although the concept of immunization had been introduced in 1796 by Edward Jenner's prevention of smallpox through inoculation with cowpox extract, a similar regimen was not possible for diphtheria. Attempts at immunization with small doses of the bacterium ended up causing disease. Another approach was needed, and it was found by German researchers who injected the toxin into a horse. This did not harm the animal, but it provoked the production of white blood cells from which an "antitoxin" was isolated. That antitoxin turned out to be the first antibody ever identified. While it saved the lives of many diphtheria victims, it did not prevent the disease, nor did it stop it from spreading.

Then in 1923, Gaston Ramon made a landmark discovery at the Pasteur Institute in Paris. Exposing diphtheria toxin to minute amounts of formaldehyde resulted in the loss of its toxicity, but the ability to stimulate antibody production was retained. This opened the door to the possibility of formulating a vaccine with the altered diphtheria toxin, now renamed a "toxoid." Indeed, by the end of the 1920s, carefully controlled trials in some 36,000 Canadian children had shown that toxoid injections reduced diphtheria incidence by at least 90 percent. The first modern vaccine was born!

Alexander Thomas Glenny, a British immunologist, was to make the next important discovery. In 1926, he noted that vaccinated guinea pigs had a better response when the injected toxoid caused a local inflammation at the site of the injection. He then had the idea of adding substances to the vaccine to cause such an inflammation and found aluminum salts to be particularly effective. He had serendipitously discovered the first vaccine adjuvant, the term deriving from the Latin "adjuvare," meaning "to help." Indeed, adjuvants help by boosting antibody production, resulting in longer-lasting immunity and a reduction in the dose of the active agent, the antigen, needed in the vaccine.

Aluminum compounds have been widely used as adjuvants in various vaccines and have a long safety record, but they do not work for all vaccines. Once the concept of adjuvants was established, the search was on for new and improved versions, several of which have been discovered. For example, the Shingrix vaccine against shingles is boosted by an ingredient found in the bark of the South American soap tree, and a flu vaccine, Fluad, specially developed for the elderly because of their poorer immune response, uses squalene derived from shark liver as an adjuvant.

Adjuvants are destined to play an important role in COVID-19 vaccines since by reducing the amount of antigen needed, they can increase the number of doses produced. Some companies are using squalene, which has led to controversy based on concerns promoted by an organization called Shark Allies that about half a million sharks will be sacrificed so that their livers can be harvested for the production of squalene. Drug producers retort that they use the livers of sharks caught for food and that they do not hunt sharks just to produce vaccines. Nevertheless, they do agree that another source of squalene is desirable.

Another source is the residue left when olives are pressed for their oil. This can be extracted to yield squalene, although the process is not as economical as isolating the chemical from shark liver. In any case, the amount of squalene used to produce vaccines is far less than that used by the cosmetics industry. Squalene is an excellent moisturizer and is found in numerous creams and lotions. For reasons of shark conservation, researchers are working on introducing genes that code for the production of squalene into microbes such as yeasts and bacteria, which will then churn out squalene, much to the relief of sharks. The Pfizer vaccine does not use an adjuvant.

Shark Allies point to another problem that plagues shark populations, namely the appetite in Asia for shark fin soup. In Chinese culture, shark fin is supposed to boost sexual potency, prevent disease, have a rejuvenating effect, and increase *chi*, a mythical form of energy. There is no basis for any of this, but the beliefs lead to the large-scale practice of finning by poachers. Live sharks are caught and their fins are cut with a hot blade before the sharks are dumped back into the water where, unable to swim properly, they soon die.

Chemical ingenuity may come to the shark's rescue in a curious fashion. Fake shark fin soup! In real shark fin soup, the fin is cut into thin strands that can be mimicked visually with noodles made from konjac root, seaweed, and the pulp from a species of squash that has appropriately been called "shark fin melon." The fake soup has the same medicinal value as the real thing. None.

THE DRY ICE STORY IS ANYTHING BUT DRY

The Phantom of the Opera is the most successful theater production of all time, playing to over 140 million people, in thirty-five countries, in 166 cities, and in fifteen languages since it opened in 1986 at Her Majesty's Theatre in London. Each performance uses 250 kilograms of dry ice to create the illusion of the Phantom's boat gliding over a misty subterranean lake, taking him and the kidnapped Christine to his lair under the Paris Opera House. That is a lot of dry ice used over thirty-five years!

Dry ice is dry, but it isn't ice. It's solid carbon dioxide, first observed in 1835 by Adrien-Jean-Pierre Thilorier, who directed a jet of the gas from a cylinder into a small vial and noted that it "quickly fills with a white, powdery, fluffy material." Thilorier had invented a machine to compress gases into liquids that he then collected in metal cylinders. One of these gases was carbon dioxide, first described by Scottish physician Joseph Black in the 1750s. Black treated limestone (calcium carbonate) with acids and captured the gas produced, calling it "fixed air." Then in 1767, Joseph Priestley used the same reaction to infuse water with this "fixed air" and produced soda water. Thilorier's compressor made the process much easier.

Thilorier's fluffy material turned out to be dry ice. Converting a liquid into a gas requires heat, and in this case, the heat is taken from the liquid carbon dioxide, cooling it and freezing it into dry ice. Today, dry ice is manufactured on a large scale by essentially the same process. The snow-like solid powder can then be compressed into small pellets or into larger blocks. Most commonly, dry ice is used to preserve ice cream, seafood, meat, and now, of course, to preserve Pfizer's mRNA vaccine, which has to be kept at –70 degrees Celsius. But let's not forget about the dry ice needed for special effects.

The principle here is straightforward. Dry ice is dumped into hot water, causing it to quickly vaporize. The vapor is still very cold and causes moisture in the air — of which there is plenty because of the hot water — to condense into droplets, essentially forming fog. Since both carbon dioxide and water droplets are heavier than air, the fog hugs the ground and creates the effect of a cloud covering the stage. The *Phantom* production uses six fogging machines consisting of heated stainless-steel drums into which dry ice is dumped. They are equipped with insulated polyvinyl chloride pipes to conduct the resulting fog to the stage.

While COVID-19 has obviously reduced the need for dry ice in theatrical productions, the demand for shipping vaccines has greatly increased and has put a strain on production. The manufacture of dry ice relies on the availability of carbon dioxide, which one might think is not an issue given all the concern about carbon dioxide levels rising in the atmosphere. Carbon dioxide could indeed be produced by liquifying air, but the atmospheric content is only about 400 parts per million. While this has been increasing, and the rise is significant when it comes to climate change, sourcing carbon dioxide from air is not economically viable.

Commercially, carbon dioxide is sourced from the industrial synthesis of ammonia or ethanol. Ammonia synthesis is one of the most important chemical reactions ever discovered and has played a significant role in feeding the world through the production of fertilizer. In the Haber-Bosch reaction, hydrogen is reacted with nitrogen at high temperature and pressure to produce ammonia, which is used as a fertilizer directly or through conversion into ammonium nitrate. To produce hydrogen, natural gas or other hydrocarbons are reacted with steam to yield carbon monoxide and hydrogen. The carbon monoxide is further reacted with steam to yield more hydrogen and carbon dioxide. This is the carbon dioxide that is sold for use in the soft drink industry, to meat production plants for stunning pigs and poultry before slaughter, and to manufacturers of dry ice. Whenever ammonia plants shut down for maintenance, or because oil prices are too high, the other industries suffer.

The ethanol industry that supplies the ethanol used as a gasoline additive is also a source of carbon dioxide. When starch or cellulose is degraded to glucose, which is then fermented with yeast to produce ethanol, carbon dioxide is a byproduct. With the decline in transportation, less ethanol is being produced, and therefore less carbon dioxide is available.

Does dry ice pose any kind of danger? You certainly don't want to touch it with bare hands because it will almost instantly freeze skin. How about the inhalation of carbon dioxide gas when dry ice sublimes? The gas is not toxic, but inhalation of a large dose can suddenly elevate blood levels and cause unconsciousness. A larger problem, though, is asphyxiation due to displacement of oxygen in the air. A terrible example of this occurred last February in Russia when, at a birthday party celebration, 25 kilograms of dry ice was tipped into a small pool in a bathhouse to create a visual effect to impress the guests. Three people in the pool suffocated as the heavier-than-air carbon dioxide displaced oxygen above the water. Due to the fog that formed, their struggles were not seen by others at the party. In the *Phantom of the Opera* production, this is not an issue since both the Phantom's and Christine's heads are well above the "swirling mist upon a vast, glassy lake." Thank goodness since the scene just precedes the Phantom launching into "The Music of the Night," the show's signature song.

THE BITTER SAGA OF QUININE

I have never been fond of drinking bitter tonic water, but there is a sweet use to which it can be put. The quinine in the water, which is responsible for the distinct bitter flavor, can make for a great demonstration of fluorescence. This is a phenomenon by which radiation of one wavelength is absorbed by a substance and is immediately reradiated at a different wavelength. The effect lasts as long as it remains exposed to the incident radiation, in contrast to phosphorescence, in which the emission continues even after the exciting radiation has been removed.

In the case of quinine, electrons in the molecule can absorb "black light," which is ultraviolet light in the 320 to 400 nanometer range, and transition to an excited state. Energized electrons then return to the relaxed state, and when they do, they emit some but not all of the energy they had absorbed. This lower-energy, higher-wavelength radiation is in the form of visible blue light. Shine black light on a glass of tonic water and it will emit a beautiful blue glow! But you can forget about tonic water shining light on a COVID-19 infection.

Why are people asking about a connection between tonic water and COVID-19 in the first place? It seems to be prompted by confusion about quinine and chloroquine, the drug being hyped for the treatment of coronavirus infection. Both quinine and chloroquine are antimalarial drugs, so maybe if chloroquine works, which has not been scientifically demonstrated, then quinine will work as well. While quinine and chloroquine have some similarity in molecular structure, they are significantly different. That difference may not matter when dealing with the malaria parasite, but dealing with a virus is a different story. Furthermore, even if quinine had some antiviral activity, the amount added to tonic water, about 80 milligrams per liter, is far too little to have an effect.

While tonic water as produced today has no medicinal effect, there was once a version that did. Let's go back to colonial India in the nineteenth century, when British soldiers were required to take quinine to prevent malaria. They were told to stir the powder in water and drink the "tonic," a term that derives from the Greek "to stretch" since the tonic was meant to stretch health. Many soldiers found the taste of quinine too bitter and did not take the medicine.

How to persuade the soldiers to take the life-saving drug? Mask the bitter taste. But with what? Juniper berries did the job nicely, and these are the berries that give gin its distinctive flavor. Soldiers didn't need much convincing to mix their tonic with gin, and they happily downed the G&T. A classic beverage was born.

Where did the idea that quinine could prevent malaria come from? For that, we take a historical trip to South America. When Jesuit missionaries arrived in the latter part of the seventeenth century to teach the natives about Christianity, they also learned something themselves. The Incas had a way of treating fever, a characteristic symptom of malaria, by drinking a brew they made from the bark of a tree. That tree was the cinchona tree, the bark of which harbors quinine, which of course was not known at the time. In any case, the Jesuits introduced the powdered bark to Europe, where it became known as Jesuit powder or the Pope's powder.

In 1820, French researchers Pierre-Joseph Pelletier and Joseph Bienaimé Caventou managed to isolate the bark's active ingredient and coined the term quinine from the Inca "quina-quina," meaning "bark of bark." Prior to 1820, the bark was first dried, ground to a fine powder, and then mixed into a liquid, usually wine, which was then consumed. Once quinine could be extracted from the bark, it became easier to determine dosages and large-scale use of quinine as a prevention for malaria began around 1850.

There was never enough quinine to meet the needs and that raised the prospect of a synthetic version. Since molecular structures could not be determined at the time, the difficulty of this task was not realized. Nevertheless, a futile effort to make quinine from coal tar in 1856 by young William Henry Perkin had a serendipitous outcome. Instead of quinine, he accidentally produced a synthetic dye, mauve, that laid the foundation not only for the commercial dye industry, but for the pharmaceutical industry as well.

The synthesis of quinine was finally achieved by Nobel Laureate Harvard chemist Robert Woodward in 1944, but it was far too complicated for commercial use. To this day, the cinchona tree remains the only source of quinine. A further problem was that with the increased use of quinine, the malaria-causing parasite was developing resistance to the drug. This stimulated a search for molecules that had some similarity in molecular structure to quinine but could be more readily synthesized.

One of these, developed by the Bayer Company in Germany in the 1930s, was chloroquine. Since chloroquine had significant side effects, a derivative with a better side-effect profile, hydroxychloroquine, was introduced by the Sterling pharmaceutical company in 1955. This is the compound that emerged from the shadows into the spotlight in 2020 when exploratory studies suggested it could be of some help in combatting the SARS-CoV-2 virus. Subsequent studies did not support the original optimism and the spotlight has dimmed, although not turned off. There are some who still maintain that hydroxychloroquine has a role to play in the COVID battle, and they are still shining a light on the compound, but it is more of a dim flashlight.

THROWING LIGHT ON ULTRAVIOLET LIGHT

"Nature and Nature's laws lay hid in night: God said, Let Newton be! and all was light." So proclaimed English poet Alexander Pope, using light metaphorically. However, Newton, the father of modern physics, was actually interested in the properties of light. In a classic experiment, he demonstrated that passing sunlight through a prism separated it into violet, indigo, blue, green, yellow, orange, and red colors. Today we recognize that these hues represent different wavelengths. The visible spectrum, the light to which our eyes are sensitive, ranges from violet with a wavelength around 400 nanometers to red at 800 nanometers. Longer wavelengths fall into the "infrared" region, which we cannot see, but can experience as heat. Shorter wavelengths, beyond the violet region, are referred to as "ultraviolet."

Like visible light, ultraviolet spans a range of wavelengths, roughly from 100 to 400 nanometers. By convention, this range is subdivided into UVA, from 315 to 400, UVB from 280 to 315, and UVC from 100 to 280 nanometers. Since wavelength is inversely proportional to the energy that a wave can carry, the shorter wavelengths are more energetic. Visible light does not have enough energy to break chemical bonds, but UVB and UVC are energetic enough to damage molecules. Luckily, the Earth's ozone layer filters out most UVC from sunlight, but UVB does get through. If skin is exposed to UVB for a sufficient time, it can disrupt bonds in DNA, sometimes resulting in skin cancer. Sunscreens and sunblocks offer protection by absorbing or reflecting ultraviolet rays.

Viruses also contain genetic material that can be attacked by ultraviolet light, with the peak germicidal efficiency being at 264 nanometers in the UVC range. Such light can be used to disinfect inanimate objects, but it cannot be used in any situation where humans are exposed because of the carcinogenic potential of this high-energy shortwave radiation. Disinfecting equipment that produces UVC is used by food manufacturers, water treatment facilities, pharmaceutical companies, hospitals, and air purification systems, but of course only when there is no possibility of human exposure. There are even robotic systems to disinfect empty busses, trains, subway cars, and operating rooms.

Concern about the SARS-CoV-2 virus surviving on surfaces, sometimes for days, raises the question of using UVC to reduce the chance of infection. Devices that are designed to disinfect small objects such as cell phones are effective and pose no risk because they are closed when operational. Effective disinfection requires both the right wavelength and the right dose of radiation. The dose in turn is a function of the length of exposure to the radiation and of its intensity at a specific distance from the source (the irradiance), usually measured in milliwatts per centimeter squared (mW/cm^2).

In an enclosed system, both the distance from the source and the irradiance can be designed to ensure disinfection. But what about the various ultraviolet-C light wands that are being promoted for the disinfection of surfaces? These are engineered to ensure that the operator is shielded from the light, which is critical. However, while the wavelength at which the device operates may be stated, the irradiance is not, so it is not possible to accurately determine the likely effectiveness. In general, disinfection requires moving the wand very slowly and very close to the target surface. Also, the light will inactivate only the viruses it directly encounters, not the ones that may be hiding in the shadows on an uneven surface. Basically, this is a more inconvenient and no more effective method of disinfection than wiping a surface with a disinfectant.

Surprisingly, within the UVC range, not all wavelengths have the same effect. Biophysicist David Brenner of Columbia University Medical Center in New York found that UVC in the 207 to 222 nanometer range can inactivate bacteria and viruses but does not harm mammals, because it is absorbed by the outer, nonliving layers of skin, preventing penetration. Brenner has shown that UVC at 222 nanometers can inactivate aerosolized H1N1 influenza virus and believes that ultraviolet lamps that emit this specific wavelength could be effective at controlling the spread of COVID-19 if installed in public places. The technology has been submitted to regulatory agencies for approval. Because of the complicated filtering system needed to produce one specific wavelength, such lamps are expensive to produce.

There is one final arrow in the UVC quiver. Back in the 1940s and '50s, ultraviolet blood irradiation (UBI) was used to treat a variety of diseases that included bacterial blood poisoning, tuberculosis, arthritis, and polio. About 7 percent of a patient's blood was withdrawn, circulated through a device where it was exposed for ten seconds to UVC at a peak wavelength of 254 nanometers, and then immediately infused back into the patient. A number of successful treatments were reported, but the technology faded into the shadows with the advent of antibiotics. Some researchers have suggested that it may now be time to take another look at the "cure that time forgot."

HYPERBARIC CHAMBERS AND COVID

It was a unique sight: a giant steel sphere, some 64 feet in diameter and five stories tall, being constructed in 1926 along Cleveland's lakefront. It was not some sort of tourist attraction; it was destined to be a "hyperbaric sanitarium" for patients with respiratory problems. And therein lies a story.

By the first decade of the twentieth century, the inhalation of nitrous oxide, or ether, for surgical anesthesia had become well established. While this was a huge advance both for patients and surgeons, it was not without problems. There were issues with asphyxia due to lack of oxygen being delivered, there were pulmonary complications, and patients often experienced postoperative nausea. Dr. Orval J. Cunningham, Professor of Surgery at the University of Kansas, sought to solve these problems by designing an anesthesia machine that allowed for careful regulation of the amount of nitrous oxide, ether, and oxygen being delivered. Cunningham was an expert in respiration, so in 1918, when the influenza pandemic began to roll across the U.S., his focus naturally shifted from anesthesia to the treatment of respiratory symptoms associated with the flu.

Cunningham had previously noted that patients with lung disease improved after moving from Denver to Kansas City. Now, when reports began to indicate that the death rate from the flu was greater in areas of higher elevation than in coastal regions, he concluded that the variable was an increase in oxygen pressure at lower altitudes. Could subjecting flu patients who were having breathing difficulties to higher oxygen pressure result in a clinically meaningful effect, Cunningham wondered? It was worth a try, but there was the question of how to go about it.

Scientists already knew that upon cooling, air condenses into a liquid from which oxygen can then be distilled and collected in cylinders. Cunningham of course had used oxygen in his anesthesia machine but thought that for the treatment of respiratory problems, oxygen delivered at a higher pressure would be more useful. This could be done with the aid of a chamber that featured above-normal atmospheric pressure through air being forced into it with the use of mechanical compressors. Air contains about 20 percent oxygen, and greater pressure exerted in such a hyperbaric chamber would force more oxygen into the lungs of anyone inside.

Various hyperbaric arrangements were already available for the treatment of decompression illness experienced by workers who had spent time in a diving bell constructing bridges underwater. The pressure inside the bell was created by pumping air from the surface and had to be high in order to keep water from entering. Inhaling pressurized air leads to more nitrogen and oxygen dissolving in the blood, and if the pressure is suddenly reduced by the bell being raised to the surface too quickly, the gases escape from the blood and form bubbles. This causes a painful condition known as the bends. The solution is a hyperbaric chamber that forces the gases back into the blood from where they can then be released more slowly.

Cunningham borrowed a small hyperbaric chamber from a local diving contractor and began to treat influenza victims. He soon reported that "patients whose lips bore the blue-black livid stamp of the kiss of death and were deeply unconscious, but if not too far from the brink, in a matter of minutes were brought back to normal color and return to consciousness."

Having had success with influenza patients, Cunningham became overly enthusiastic and began to experiment with hyperbaric treatment of a host of other conditions such as hypertension, syphilis and diabetes. Although he had no significant evidence of any benefit, he managed to attract patients desperate for help. One of these was Henry Timken, an industrial magnate who had made a fortune manufacturing bearings. Timken suffered from kidney problems that apparently improved with Cunningham's treatment. He was so elated that in 1926 he donated $1.5 million to the doctor for the construction of that 64-foot diameter, five-storey steel sphere in Cleveland that would serve as a "hyperbaric sanitarium." Patients were housed in one of the thirty-eight rooms and subjected to air pumped in at high pressure.

Cunningham's claims did not sit well with Morris Fishbein, head of the American Medical Association, who opined that such treatments seemed "tinctured more strongly with economics than with scientific medicine." This led to a decline in clientele, and Cunningham eventually sold the facility to buyers who converted it into a general hospital. This too had financial problems and was closed in 1937. Eventually, the giant steel sphere was razed in 1942, and the metal was used for the war effort.

Today, COVID-19 victims have access to equipment that can deliver pure oxygen through masks or tubes at doses higher than were available in Cunningham's compressed air chamber. Also available today are hyperbaric oxygen chambers that provide not air but pure oxygen at high pressure. Apparently, these are being tried on patients in China with some optimistic results being reported.

It seems that hyperbaric oxygen can penetrate inflammatory pulmonary secretions, allowing adequate oxygen to reach the blood while simultaneously inhibiting the inflammatory process. Who knows, we may be building giant hyperbaric spheres again.

CORONA AEROSOLS

Molecules are very small. So small that every breath we take contains about 2.4 x 1,022 of them. The vast majority of these are oxygen and nitrogen, with lesser amounts of argon and carbon dioxide. The same goes for every breath exhaled. When random mixing in the atmosphere is accounted for, we can calculate that every breath we take contains some molecules that have once been exhaled by Albert Einstein, Napoleon, or Jesus.

That's all very interesting, but the fact that every breath we take has at least five molecules that appeared in Abraham Lincoln's last breath has no practical significance. However, the number of coronavirus particles that are expelled in every breath of the asymptomatic infected stranger with whom we may share airspace in a room, bus, airplane, or car *is* of significance.

Unfortunately, as with almost every other aspect of the SARS-CoV-2 virus, we have very limited data. A guesstimate can be made based on some mouse experiments with the original SARS virus. The infective dose in that case turned out to be only a few hundred viral particles, which would suggest that humans could be infected by as little as a few thousand inhaled particles, or virions, although we have to keep in mind one of the dictums of virology: "mice lie and monkeys exaggerate."

The problem is that even if we accept the infective dose to be possibly as little as a thousand virions, we don't have a real grasp of how many SARS-CoV-2 particles are released by an infected person's cough, sneeze, or breath. However, based on some experiments with the influenza virus, the number of particles in the droplets produced by a sneeze or cough may be as many as 200 million. Obviously then, being directly in the path of a cough or sneeze can cause an infection. As these droplets spread, ones larger than 5 microns fall down due to gravity, but the smaller ones, less than 5 microns, termed aerosols, continue to drift farther and can be detected at a distance greater than 2 meters. These tiny particles can stay suspended in the air for hours and because of their size can be inhaled deep into the lungs, bypassing an encounter with immune cells higher up the respiratory tract.

True, by this time the concentration of virions per unit volume may be very low, but if someone spends enough time in this environment they can still become infected. That's because the total number of particles to which one is exposed is determined by the number of breaths taken times the number of particles per breath. Walking by someone on the street who may be exhaling viral particles, even if they are jogging and breathing heavily, presents very little risk because the time spent inhaling those particles is very brief, making it unlikely for an infective dose to be reached. On the other hand, being in the same room with someone shedding virus can be an issue even at a significant distance if the exposure time is long enough.

This was dramatically demonstrated in one of the most widely investigated cases of transmission, the now-famous choir practice in the town of Mount Vernon, in Washington State. In spite of distancing and avoiding any sort of sharing, a single asymptomatic carrier managed to infect forty-five of sixty people. Many of these never came anywhere close to the infected person, so by the time the viral particles reached them, they had been significantly diluted. But the practice went on for over two hours, so although each inhaled breath contained few viral particles, there were a lot of breaths taken.

This is also the reason why meatpacking plants, telephone call answering centers, religious institutions, prisons, and seniors' residences have become hotbeds for disease. Lots of people indoors sharing the same airspace. Furthermore, singing, as well as yelling and even loud talking, increases the number of droplets and the distance they are expelled. In a meatpacking plant, for example, the workers have to speak very loud over the din of the machinery.

Where does this leave us? Not in a very comforting situation. Under many indoor conditions, aerosols can remain airborne for hours and drift farther than the 2-meter "safe" distance. If an extended period of time is spent in an indoor environment that is occupied by more than a few people, a facemask can reduce risk. This would be more important for workers in a store than for customers. Restaurants, unfortunately, pose a problem because of the time spent dining.

Searching for somewhat of a bright note on which to end, research has shown that on the *Diamond Princess*, the cruise ship that was quarantined for two weeks, transmission of the virus was through close personal contact. No transmission was detected through the ship's ventilation system.

Eventually, we will get back to some degree of normalcy. But don't hold your breath. Go ahead and take one, remembering that it will contain some molecules from the perfume used by Marilyn Monroe, from the aroma of a steak cooked by Julia Child, and from the sweat of Donald Trump. And undoubtedly some SARS-CoV-2 particles.

THE FASCINATING HISTORY OF DEXAMETHASONE

Put that one on the front burner, said the National Research Council in the U.S. in 1941. That "one" was the isolation and purification of the substance produced by the adrenal glands that allow the body to cope with stress. Surprisingly, this project was given higher priority than the development of antimalarial drugs and penicillin. The need for immediacy was generated by intelligence reports that the Nazis were importing bovine adrenal glands from Argentina by submarine, supposedly to extract a substance that would allow the Luftwaffe pilots to handle the stress of reduced oxygen in the blood when flying at high altitudes. The story of importing adrenal glands was never confirmed, and there is no evidence the Nazis ever used any sort of adrenal extract.

While the Nazis may not have imported bovine adrenals, the possibility of using adrenal extracts to improve human health and performance was not farfetched. In 1855, British physician Thomas Addison published a landmark monograph linking pathological changes in the adrenal glands to a disease in patients who exhibited a particular kind of skin discoloration, "feebleness of the heart's action," and "general languor and debility." That disease, characterized by poorly functioning adrenal glands, came to be known as Addison's disease.

In the 1930s, Edward Calvin Kendall, professor of physiological chemistry at the Mayo Clinic in Rochester, Minnesota, was intrigued by the adrenal glands and sought to find the substance they produced that allowed animals to survive once their adrenals were removed. The putative substance at the time was referred to as "cortin." In his search for this elusive substance, Kendall extracted thousands of pounds of animal adrenals and found that there was no one cortin, rather the adrenals produced a number of compounds. By 1940, through studies with animals whose adrenals had been removed, he concluded that one of these compounds, called compound E at the time, had the greatest activity in prolonging the animals' lives and was a candidate for the treatment of Addison's disease. He also noted that without adrenal glands, the smallest physical stress would precipitate the demise of the animals.

Kendall's adrenal research at the Mayo Clinic came to the attention of physician Philip Hench, a Mayo colleague, who had made a remarkable observation. Patients who suffered from arthritis improved with the onset of jaundice. Hench also noted that pregnancy relieved arthritic symptoms and postulated that under these conditions, the body produces some anti-inflammatory substance. He wasn't a chemist, so he sought help from Kendall, who suggested that bile acids (which build up in jaundice) and as pregnancy hormones were chemically similar to his compound E, which upon identification of its molecular structure was renamed cortisone since it was extracted from the outer layer, or cortex, of the adrenal glands. By 1948, Kendall, with help from Lewis Sarett at Merck, had developed a synthetic method to produce cortisone from ox bile. Now having significant amounts at his disposal, he suggested that Hench try it on one of his arthritis patients.

An opportunity presented itself when one such patient at the Mayo Clinic refused to leave until she felt better. "Mrs. G" was treated with cortisone, and within days showed almost miraculous improvement. Her pain resolved to such an extent that she went shopping and claimed: "I have never felt better in my life." Fourteen other patients suffering from arthritis were soon treated with similar results. The word was out about a miracle drug, but availability was scarce, giving rise to a black market for fake cortisone. But by 1952, a method to produce the drug from diosgenin, a compound found in a variety of plants, had been developed and cortisone began to perform its magic on arthritis sufferers.

The miracle, however, was tainted by side effects with prolonged use. Patients suffered from periods of depression, mood changes, fluid retention, thinning of the skin, and weakening of bones. The search was on to find synthetic analogues with increased efficacy and fewer side effects. Prednisone turned out to be more active and had fewer side effects. It was used to treat President Kennedy's Addison's disease.

In 1958, both Merck and Schering introduced dexamethasone, which was far more potent than cortisone and had a longer duration. But like all corticosteroids, which encompass compounds produced by the adrenal cortex and their synthetic analogues, dexamethasone has significant chronic effects that include osteoporosis, weight gain, changes in fat deposition, increased risk of ulcers, and immune reaction suppression. Nevertheless, it is a very effective drug in the treatment of chronic obstructive lung disease, severe allergies, asthma, swelling of the brain, autoimmune conditions such as arthritis, and now in COVID-19.

Like any drug, dexamethasone can be abused. One example is its widespread use by sex workers in Bangladesh. While brothels are legal in many areas of the country, there is very little monitoring of the age of the women, many of whom are below eighteen. Competition for customers is keen, and since the men apparently prefer larger and more rounded bodies, the women ply themselves with dexamethasone, which is cheap and available over-the-counter. It also provides more energy, which is welcome since they service up to fifteen men a day. It is a sorry situation.

Dexamethasone obviously has a fascinating history and possibly an important role in the future. A British trial has already shown that patients given the drug while on ventilators suffer one-third fewer deaths than control subjects. Those requiring oxygen without a ventilator have 20 percent fewer deaths. Perhaps dexamethasone may even work on patients who are at an earlier stage of the disease. That is why President Donald Trump was given dexamethasone, even though that treatment is not in step with current guidelines.

BRAZILIAN PIT VIPERS, BLOOD PRESSURE, AND COVID-19

You would not want to meet a Brazilian pit viper in the jungles of South America. No, these snakes that can grow up to five feet in length do not live in pits. Their name comes from the heat-sensing glands found on either side of their triangular-shaped head that look like little pits. They enable the snake to locate prey in total darkness! The viper's venom is so potent that Indigenous people have used it to tip their poison arrows. But it was not only people in the jungle who were interested in the pit viper's venom. Scientists in the lab were also intrigued by the poison's mechanism of action. They knew that banana plantation workers who were bitten by the snake quickly collapsed from a drop in blood pressure.

In 1939, Dr. Mauricio Rocha e Silva became intrigued by this effect and injected pit viper venom into rodents to investigate what sort of changes it would cause in their blood chemistry. It took a few years before a peptide that was to be named bradykinin, from the Greek for "slow" and "movement," was isolated from the animals' bloodstream. The subjects moved slowly alright, and eventually moved not at all. Bradykinin caused a dramatic drop in blood pressure that often led to death. Not always, though, since the body recognizes bradykinin as a foreign substance and mobilizes an enzyme that can break it down. So, unsurprisingly, dosage of the venom is critical. A small dose can be survived, but a larger dose is too much for the enzyme to deal with. In any case, the action of bradykinin introduced the possibility of using this chemical as a blood pressure lowering drug. However, since bradykinin is a peptide, a short chain of amino acids, it cannot be taken orally. It is broken down during digestion and doesn't enter the bloodstream.

All was not lost, however, thanks to research in the 1960s by Dr. John Vane and colleagues at the University of London that revealed a mechanism by which the body raises blood pressure when that is required. A substance called angiotensin forms upon a signal from the kidneys and is then converted into angiotensin II, a peptide that constricts blood vessels and results in an increase in blood pressure. This conversion requires an enzyme, appropriately named angiotensin converting enzyme, or ACE. In 1968, Vane, who would go on to win the 1982 Nobel Prize for physiology and medicine for the discovery of the mechanism of action of aspirin, showed that viper venom peptides inhibit the activity of this enzyme and prevent the formation of angiotensin II, thereby lowering blood pressure. If scientists could figure out what part of the bradykinin molecule binds to the enzyme to inhibit it, a drug based on that part of bradykinin's molecular structure could possibly be developed to lower blood pressure.

Between 1970 and 1973, chemists at ER Squibb & Sons tested over 2,000 different compounds that had molecular structures similar to parts of bradykinin and were finally rewarded by finding the first orally effective ACE inhibitor. It would hit the market in 1981 as captopril. Eventually, a number of other ACE inhibitors such as enalapril (Vasotec) and ramipril (Altace) were developed with a better side effect profile than captopril, and these have found widespread use not only for lowering blood pressure, but also in the treatment of congestive heart failure and kidney problems.

The scourge of COVID-19 and insight into how the virus infects cells raised an issue about the use of ACE inhibitors. When these drugs are administered, the body senses the drop in blood pressure and kicks in some help with the formation of another enzyme, called ACE2, that reconverts the pressure-elevating angiotensin II back into angiotensin. Now here is the problem. This enzyme, ACE2, can also end up attached to the surface of cells, where it acts as a handle for the SARS-CoV-2 virus. The gripping of this handle is the first step in the virus's entry into cells, where it then goes on to replicate.

An obvious question then is whether people taking ACE inhibitors are at greater risk for COVID-19 since with an increase in ACE2, the virus has more entry points into cells. An important question since these drugs are widely used. Early observations indicated that people who were taking ACE inhibitors suffered more serious COVID cases if infected, but follow-up studies with control groups did not find a link to these drugs. At first it seemed like these people were at greater risk because people who take ACE inhibitors take them for hypertension or heart problems, and these conditions predispose them to COVID-19.

Nevertheless, upon reading the early reports, some people gave up their ACE inhibitors. That is a problem unless they are replaced with other blood pressure lowering drugs, which are available. Being bitten by a pit viper would not be a good alternative. It is interesting to note that the original symbol for medicine, "The Rod of Asclepius," is a serpent-entwined rod wielded by the Greek god of healing and medicine, Asclepius. Since snakes were immune to their own poison, they were thought to have magical healing properties. Not totally wrong. Maybe they're not quite magical, but ACE inhibitors developed from research into snake venom have had a huge impact on the treatment of hypertension and congestive heart failure.

ZAPPING COVID WITH COPPER?

Any accomplished chef will sing the praises of copper cookware. The metal spreads heat evenly, no hot spots to worry about. Egg whites whip better in a copper bowl, an effect I have actually confirmed by putting it to a test. A "Moscow mule" tastes better when consumed from the traditional copper mug, although admittedly this may be purely psychological. I have even investigated the supposed health benefits of copper bracelets and found nothing in the scientific literature in terms of any supporting evidence. But I have never thought of inserting a copper rod up my nose to prevent a bacterial or viral infection. That's because I had never before heard of the CopperZap, or its claim that it can zap bacteria and viruses out of existence.

The CopperZap is just a chunk of copper with one end wide enough to serve as a handle, while the other narrows into a shaft thin enough for insertion into the nose. Massaging the nostrils for 60 seconds with metallic copper can supposedly zap the living daylights out of any bacteria or viruses that have infiltrated the nasal tissues. Furthermore, small amounts of copper ions will be left behind to offer a protective effect. An added bonus is inactivation of any germs that may be present on the hands. At least so goes the claim.

When this zapper was first marketed, it targeted the rhinoviruses that cause colds. But bloggers then chimed in with claims of successfully treating cold sores, flu, sinus problems, nighttime stuffiness, and even illness after airline travel. With COVID-19, bloggers began to hype it as protection against the disease, although the manufacturer did not make such a claim. For whatever reason, the CopperZap website has been stripped of all information except the message that the device is no longer available.

At first glance, poking the nose with copper for protection against infection sounds like one of the bevy of snake oil claims that flood social media. But there may actually be something to this, although not much. What we have here is the usual technique used by clever marketers of blowing a kernel of legitimate science out of proportion. Copper does have antimicrobial properties, something that was discovered empirically long before microbes were ever known to play a role in disease. *The Smith Papyrus*, an ancient Egyptian medical text that dates to somewhere between 2600 and 2200 BC, describes treating chest wounds with copper. Hippocrates recommended copper as a treatment for leg ulcers, and the Greeks sprinkled copper oxide on fresh wounds. The ancient Phoenicians reportedly inserted shavings from their bronze swords into battle wounds, bronze being an alloy of copper and tin.

Today, the antimicrobial properties of copper are well established. Elemental copper has a single electron in an outer orbit that is easily removed, leaving behind positively charged copper ions. These can throw a wrench into the multiplication of bacteria and viruses by interfering with the structure of their genetic material. Furthermore, the lost electron can be absorbed by oxygen to form superoxide, a highly reactive species that can also destroy microbes. This is why copper surfaces, such as on IV posts, armrests, bed frames, and over-the-bed tables in hospitals have been shown to reduce hospital-acquired infections.

While these surfaces do destroy microbes, they do not do so instantly. Significant contact time is required. It is most unlikely that a sixty-second massage of the nose tissues will do anything, but without this having been studied, it cannot be totally ruled out. So far, though, no evidence by the manufacturer has been provided. In any case, the $69.95 charge for the copper wand is exorbitant.

How about face masks that are formulated with copper? Dozens of these have appeared, especially since an article in the *New England Journal of Medicine* demonstrated that the SARS-CoV-2 virus survived for only four hours on a copper surface, while it remained infective on paper, plastic, and steel for days. Once again, in theory, there may be some benefit. While there is no evidence that copper in a mask increases the filtering ability, any viral particles that come out of the mouth or nose of an infected person, or particles that land on the mask from another infected person, can be inactivated by exposure to copper. That inactivation, though, takes time. Exactly how long depends on the specifics of the mask, such as pore size and whether the copper is in the form of fibers, nanoparticles, or copper compounds. So far, no information about the longevity of viruses on masks is available. Where then is the benefit?

Suppose that someone touches an infected surface such as a doorknob, and then adjusts their mask, the mask can become contaminated. Now we have a possible scenario for infection if later the mask is removed without taking care not to touch the outside, followed by touching the face. The theory is that viruses will linger for a shorter time on a mask that contains some form of copper than on a cloth mask and therefore there is a reduced chance of infection. Maybe. But before spending up to $70 for a copper mask it would be nice to have some evidence of efficacy. Same goes for massaging your nostrils with a copper gizmo.

NONSENSE WAS QUICK TO EMERGE AS VIRUS GRIPPED THE WORLD

The Earth is flat. Spoons can be bent with the power of the mind. The sun goes around the Earth. Self-levitation can be achieved by mastering meditation. A surprising number of people hold such beliefs, and while irritating to those of us who are guided by science, they are essentially benign, unlikely to harm anyone. Unfortunately, there are other nonsensical beliefs, often emerging at a time when people face desperate situations, that are not harmless. The COVID-19 pandemic that has descended on us is such a time. The SARS-CoV-2 virus can kill, but tragically some false beliefs about avoiding infection can also be lethal. A dreadful example is the death of some 500 people and the sickening of 3,000 others in Iran by drinking methanol.

How does such a calamity happen? Apparently, it all started with a post on social media by Connor Reed, an English teacher in Wuhan who presented at the local hospital complaining of a persistent cough and breathing difficulties. He was diagnosed with COVID-19 and was hospitalized for two weeks, his breathing facilitated by oxygen support. While in the hospital, he also started drinking hot whiskey with honey until it ran out, although it isn't clear how he managed to avail himself of this old-fashioned "remedy." In any case, he recovered and claimed in his post that "I am proof that coronavirus can be beaten." Given that he really did recover, technically the statement is correct. But it is quite a stretch to suggest that the recovery was due to the whiskey and honey. Reed is 25 years old, and most people of that age beat the virus whether they ply themselves with whiskey and honey or not.

Combine this post with the legitimate advice to use alcohol-based hand sanitizers, the illegality of drinking alcohol in Iran, and a distrust in a government that at first minimized the coronavirus, and you have the makings of a tragedy. The term "alcohol" refers to a family of compounds with molecular structures that feature the presence of one or more hydroxyl groups (OH) bonded to a carbon framework. However, different alcohols can have dramatically different properties. Ethanol, the alcohol in beverages, has two carbon atoms in its structure, while methanol is a simpler compound having only one carbon to which the hydroxyl group is attached. That extra carbon atom in ethanol makes a world of difference. While ethanol can be toxic in large doses, methanol is poisonous in very small amounts.

Both ethanol and methanol can depress the central nervous system, causing respiratory problems and a decreased heart rate, but methanol does it more effectively. Furthermore, methanol is metabolized first to formaldehyde, then to formic acid, which is highly toxic due to inhibition of the enzyme cytochrome c oxidase. This enzyme is needed for the proper use of oxygen by cells, and when it becomes dysfunctional, the resulting lack of oxygen, or hypoxia, can lead to death. If not death, methanol poisoning can cause blindness.

Ethanol production is illegal in Iran, punishable by lashing. On the other hand, methanol can be produced because it as an important commercial chemical used in the production of numerous substances, ranging from fuel additives to adhesives and plastics. By law, in Iran, a dye has to be added to commercial methanol to prevent anyone mistaking it for ethanol and drinking it. As is often the case where ethanol is illegal, bootleggers surface to fill the need. Some actually produce ethanol by secretly fermenting raisins, but others add bleach to destroy the coloring that is added to commercial methanol and then pass it off as drinkable alcohol. Like ethanol, methanol does cause inebriation, but that can be quickly followed by death. It seems that desperate people in Iran bought into the idea that saturating themselves with alcohol can prevent COVID-19 infection and ended up poisoning themselves with methanol. Some reports suggest that in a couple of provinces, drinking methanol has caused more deaths than the coronavirus.

Fortunately, we have not seen any such cases in North America, although people have wondered about using windshield washer fluid as a disinfectant when nothing else is available. A very bad idea! First the concentration of methanol is in the range of 40 to 50 percent, which is not enough to act as a disinfectant, and on top of that, methanol readily crosses the skin and ends up in the bloodstream. Just handling methanol can cause a toxic reaction. There is no situation in which methanol should be used to combat COVID-19.

Neither is there any situation in which people should be consuming chlorine dioxide bleach. Social media sites crow about Miracle Mineral Solution, a concoction that comes in two bottles, one containing sodium chlorite, the other citric acid. When mixed, these combine to produce chlorine dioxide, an industrial bleaching agent that is one of the approved antiviral chemicals to clean surfaces. But ingesting a chlorine dioxide solution with hopes of killing a virus is pure folly and can result in nausea, diarrhea, severe vomiting, and life-threatening low blood pressure. Miracle Mineral Solution is not a miracle, not a mineral and certainly not any solution to the coronavirus curse.

Let's move on to Dr. Sarah Myhill, a general practitioner in Wales who has put out a video about protecting oneself from infections.

She doesn't mention coronavirus specifically but does speak of viral infections in general. The date of the video suggests that it is intended to target the virus with which the world is struggling. Let me mention that Dr. Myhill has had numerous run-ins with the General Medical Council in Britain and complaints against her have resulted in some thirty investigations. Most of the complaints were about recommendations on her website that lacked evidence and her unconventional treatment of chronic fatigue syndrome with various supplements that she sells. On several occasions she was either suspended from the Medical Register or forced to practice medicine under severe restrictions.

Dr. Myhill's answer to infections is based on vitamin C, iodine, and a "salt pipe." In her own words: "Vitamin C contact kills all microbes, and we can take that from the inside out. Iodine in contact kills all microbes, and we can use that from the outside in and you use both tools to prevent infection." Then she goes on: "I liken it to a defense with Muhammad Ali, a right hook and a left punch and you get the microwave in between with a pincer movement." I could come up with a punch line here, but it wouldn't be elegant.

It is simply not true that contact with vitamin C kills all microbes, and there is no evidence that inhaling iodine vapor from a salt pipe has any effect on reducing coronavirus infections. The doses of vitamin C she recommends are astounding. She says everyone should be taking 5 grams of vitamin C a day, which is roughly a hundred times the recommended daily allowance, and two and a half times the tolerable daily intake. Then at the first sign of infection, this should be increased to 10 grams an hour until you get diarrhea. She tells her patients, "Don't wait for an infection to come along. Put the defenses in place now, so eat a 'paleoketogenic' diet. Don't take drugs." I suppose vitamin C and iodine don't count as drugs. Needless to say, there is no evidence that a paleoketogenic diet can ward off viral diseases. And diarrhea is not desirable. I wonder if her patients are hoarding toilet paper.

Then there was that widely circulating video by Dr. Dan Lee Dimke, who took a different approach. Instead of sniffing iodine, he recommended using a hairdryer to blow hot air up the nose through a mist of water. The idea is to inactivate viruses in the nose with heat. Dimke's PhD is in "psycholinguistics," which apparently qualifies him as an expert in virology. His idea of inactivating viruses in this fashion is outlined in his book *Conquer the Common Cold and Flu*. Dimke's other books include *Million Dollar Sales Letters, Meltdown Diet, Power through Persuasion*, and *Multilevel Marketing Power Pack*. A true talent this man is, judging by his bio, in which he says he started to teach college when he was seventeen and learned Mandarin in 18 days. He also reads 25,000 words a minute and plays over 20 musical instruments. Could such a prodigy possibly be wrong? Or is the World Health Organization blowing hot air when it says that it is useless to attack any virus that may be present in the nose with a hairdryer?

There is also a "warning" making the rounds that hand sanitizer "has the same ingredients as antifreeze" and that you should not let dogs lick your hands. This warning is based on chemical ignorance. The reference is to propylene glycol which is used here as a humectant, meaning it holds in moisture to prevent the hands from drying out. Propylene glycol is indeed found in some antifreeze, particularly in products that are promoted as being environmentally friendly and having low toxicity. It replaces ethylene glycol, the usual active ingredient in antifreeze, which is indeed toxic to dogs as well as humans. But propylene glycol is not toxic. That extra carbon atom in propylene glycol makes all the difference! So, there is no worry for your dog if he licks your hand after you have used a hand sanitizer.

SEINFELD AND THE CORONAVIRUS

What is the connection between *Seinfeld* and COVID-19? Believe it or not, there is one. And it has to do with perhaps the most clever of all

Seinfeld episodes, the one entitled "The Contest." Although the activity that was the subject of the episode was never mentioned by name, it was clear that it was all about Jerry, Kramer, George, and Elaine taking matters into their own hands, as it were. In some circles, this "solitary pursuit" is referred to as "onanism," a term that first appeared in the early eighteenth century with an anonymously published pamphlet called *Onania*. That title derives from the biblical story of Onan, who got into trouble when he disobeyed a direct order from God to father a child by his brother's widow. As described in the book of Genesis, instead of impregnating his sister-in-law he "spilled his seed on the ground." This displeased the Lord and cost Onan his life.

Although the most likely interpretation of "spilling his seed" would seem to be coitus interruptus, in *Onania* it was interpreted as what one might call self-pleasuring and described as a heinous sin. The pamphlet described the "frightful consequences" of "self-pollution" and offered spiritual and physical advice to those "who have already injured themselves by this abominable practice." The injuries described were frightful. Epilepsy, mental illness, tuberculosis, pimples, and even blindness were said to be possible outcomes, with the latter giving rise to the old joke, "Let's just do it until we need glasses."

Besides the supposed biblical admonition, the proposed rationale for the injurious effects was that a sexual act was followed by a sudden feeling of lethargy, and if performed too frequently would leave the body in a diseased state. Sex was to be reserved for procreation. Sylvester Graham, the nineteenth-century health guru after whom the Graham cracker is named, claimed that the loss of an ounce of semen was equivalent to the loss of four ounces of blood, although there was no explanation about how he came to this conclusion. He saw "self-abuse" as the "most criminal, most pernicious, most unnatural" of sexual acts and labeled it a contagious disease that reduced life force and exposed the body to disease and even death.

Dr. John Harvey Kellogg, who rose to fame as the founder of the famed Battle Creek Sanitarium, where the rich and famous came to be treated for diseases they probably never had, was a devotee of Sylvester Graham and had even more extreme views about sexual activity. He encouraged strict abstention and never consummated his own marriage, although he and his wife adopted close to forty children. He called "solitary vice" the most dangerous of all sexual abuses and claimed that there was a connection between food and drink and one's urges and thoughts. "Irritating foods" such as meat and refined grains stimulate desires, he said. "A man that lives on pork, fine-flour bread, rich pies and cakes, and condiments, drinks tea and coffee, and uses tobacco might as well try to fly as to be chaste in thought," Dr. Kellogg claimed.

Although John Harvey, together with his brother Will, did invent corn flakes, the popular notion that the cereal was promoted as a means of controlling one's urges is a myth. The brothers were basically interested in creating an easy-to-digest, healthy, pre-prepared breakfast, although cereal did fit into John Harvey's "unstimulating" dietary scheme:

- Never overeat. ("Gluttony is fatal to chastity and overeating will be certain to cause emissions, with other evils, in one whose organs are weakened by abuse.")
- Eat only twice each day. ("If the stomach contains undigested food, the sleep will be disturbed, dreams will be more abundant, and emissions will be frequent.")
- Don't eat stimulating food. ("Spices, pepper, ginger, mustard, cinnamon, cloves, essences, all condiments, pickles.")
- Don't drink stimulating drinks. ("Wine, beer, tea and coffee should be taken under no circumstances. The influence of coffee in stimulating the genital organs is notorious. Chocolate should be discarded. Tobacco, another stimulant, although not a drink, should be totally abandoned at once.")
- Eat and drink plain and bland foods and drinks. ("Eat fruits, grains, milk, and vegetables; they are wholesome and unstimulating. Graham flour, oatmeal, and ripe fruit are

> the indispensables of a diet for those who are suffering from sexual excesses.")

Now we come to the connection with COVID-19. Kellogg would certainly have been against the suggestion seen in some blog posts these days that practicing onanism may be helpful in preventing the disease. What is the rationale for this advice? A 2004 study of eleven male volunteers whose blood levels of natural killer cells increased after self-manipulation. These killer cells target other cells that have been infected by a carcinogen or by a virus. When a cell infected by a virus dies, the virus is also destroyed. However, there is no evidence that ipsism has any protective effect against infection by the SARS-CoV-2 virus.

There is one way that the practice might offer protection. After all, it does conform to the advice for physical distancing. Accordingly, the New York City Department of Health and Mental Hygiene recently reminded its Twitter followers that, in the age of COVID-19, "you are your safest sex partner."

PART II

THERE IS SCIENCE BEYOND COVID

CHEMISTRY — THE DR. JEKYLL AND MR. HYDE SCIENCE

It shouldn't be too surprising that *The Strange Case of Dr. Jekyll and Mr. Hyde* by Robert Louis Stevenson is one of my favorite stories. After all, it is about the link between chemistry and good and evil. Dr. Jekyll is a respected physician who discovers a chemical formula that releases his alternative personality, the evil Mr. Hyde. He also manages to formulate an antidote that turns the vile Hyde back into the good Dr. Jekyll. There are various allegorical interpretations possible here, the most popular being the human tendency both for good and evil. However, I favor another interpretation. Chemistry can be used both for good, symbolized by the antidote, as well as evil, represented by the potion that transforms Dr. Jekyll into Mr. Hyde.

Fictional stories often have roots in reality. Stevenson may well have been inspired by the strange case of Eugene Chantrelle, a French teacher in Edinburgh who had married a sixteen-year-old student after getting her pregnant. The relationship turned sour and ended with the sudden death of the young woman. Since Eugene had taken out a large life insurance policy on his wife, he became an immediate suspect and was arrested when traces of opium were found in her vomitus. Stevenson was shocked by the allegation of murder, since he had been friendly with Chantrelle, who seemingly led a completely normal life. He attended the whole trial, in which the possibility emerged that Chantrelle had also dispensed with others by serving them "toasted cheese and opium" at parties he hosted. The jury needed only an hour to set up a date with the hangman. Eugene's dual personality was a shock to Stevenson and sensitized him to the evil that may be hidden by outward appearances.

Chantrelle was not the only contributor to *The Strange Case of Dr. Jekyll and Mr. Hyde*. The famous eighteenth-century physician John Hunter, who pioneered Caesarian sections, cardiopulmonary resuscitation, and the staging of cancer, was probably best known for his contributions to anatomy. Bodies to dissect were not easy to come by and anatomists often worked closely with body snatchers and grave robbers. Since this was a clandestine activity, Hunter purchased a large mansion that had an elegant entrance for distinguished visitors in the front and a back door where the grave robbers would deliver the bodies. Stevenson was aware of this practice. His short story, "The Body Snatcher," was published in 1884, two years before his Dr. Jekyll and Hyde classic. The description of Dr. Jekyll's home, with its dual entrances, is strikingly similar to the well-documented layout of Hunter's house in London.

Why would Stevenson have chosen Hunter's house as a model? He may have been taken by the idea that the doctor who greeted eminent patients such as Benjamin Franklin and Franz Joseph Haydn at the front door cavorted with the criminal element at the rear entrance. He may also have known about some of Hunter's unusual experiments, such as the transplantation of the testes of roosters into the abdomen of hens, apparently to see if the hens would be impregnated or perhaps be transformed into cocks. Neither happened. But for Stevenson, such experiments may have been stretching the boundaries of science too far.

Some eighty years later, Hunter's work came to the attention of German physician Arnold Adolf Berthold, who extended it and essentially laid the foundations of the science of endocrinology, the study of hormones. In 1849, Berthold described removing the testes from six male chickens, two of which would serve as controls. These birds had the combs on their head collapse, their crowing falter, and their interest in hens wane. In two other birds, he implanted their own testes back into their abdominal cavity and saw them develop normally. The last pair of roosters also had testes implanted, but this time instead of their own, they received their partner's. Not only did the implanted testes result in the growth of normal combs, they caused the recipients to chase hens around with considerably more spirit.

Berthold's simple interpretation was that the body was made up of parts, all of which are needed for proper functioning. He did not recognize that the testes may have been releasing chemicals into the bloodstream that stimulated activity elsewhere. Without realizing it, he had demonstrated the effect of hormones, a term coined in 1902 (from the Greek for "stir into action") by English physiologists Bayliss and Starling after they managed to isolate secretin, a protein released by the small intestine to stimulate pancreatic secretion.

Somewhat surprisingly, there were no follow-ups to Berthold's pioneering work until 1910 when Eugen Steinach in Vienna replicated the experiment in rats and demonstrated rejuvenation with implanted testes. One of his disciples, Serge Voronoff, hypothesized that what works in rats may work in men. Given that human testicle donors were hard to come by, he resorted to our closest relatives, chimps. Hundreds of men around the world received implanted animal testes with hopes of once again being able to chase chicks like Berthold's young cocks. But in 1927 the Royal Society of Medicine stepped in and declared such transplants to be "poppycock." Still, the infatuation with these rejuvenation schemes did pave the way for research that resulted in the isolation and subsequent synthesis of testosterone, the main male sex hormone, now available to treat men with low levels. Unfortunately, it has also been abused by athletes looking to build muscle. Talk about a Dr. Jekyll and Mr. Hyde substance!

GOOD SCIENCE IS NO LAUGHING MATTER

"I was very sick and mainstream medicine let me down," my correspondent informed me. "Then out of desperation I tried the 'Miracle Mineral Solution' and it cured me in two weeks." When I responded that I didn't think that this mixture of citric acid and sodium chlorite, a common bleaching agent, was supported by any evidence, he accused me of being an industry shill and told me that Jim Humble, the "inventor" of this cure-all, was deserving of a Nobel Prize. He finished by telling me to look at all the scientists who were first ridiculed but then revered after they were proven to be correct. The inference was that Humble, who is hardly humble when it comes to making claims, would eventually be recognized as a modern-day Galileo.

It is certainly true that there have been scientists and physicians who at first were ridiculed and were subsequently recognized as visionaries. Galileo is often presented as an example of someone who was first castigated because of his support of Copernicus's theory that the Earth orbits the sun and was later rehabilitated as the "father of modern science." Indeed, the Italian polymath was chastised, but not because his science was shown to be incorrect, but because at the time the Catholic Church held the biblical view that the Earth was the center of the universe. When evidence mounted that Galileo was correct, the Church acquiesced, although it took 350 years for Pope John Paul II to declare that Galileo had been "imprudently opposed."

There are numerous other examples of scientific ideas that were first opposed then embraced. Edward Jenner, who long before vaccines were accepted medicine used cowpox lesions to successfully inoculate for smallpox, and Ignaz Semmeleis, whose promotion of handwashing among attendants in obstetrics wards saved the lives of many new mothers, were both ridiculed initially for their theories.

A more recent case is Australian gastroenterologist Barry Marshall's contention in the 1980s that ulcers are caused not by stress, spicy foods, or hyperacidity, but by a bacterium. This idea was widely derided until Marshall, in a somewhat foolhardy experiment, swallowed a generous dose of *Helicobacter pylori* bacteria and developed gastric problems that were cured with antibiotics. This convinced researchers to organize proper trials, and before long antibiotic treatment of ulcers became common practice. In 2005, Marshall and his collaborator Robin Warren received the Nobel Prize for their work.

William Harvey's concept of blood circulation, Gregor Mendel's theories about the laws of inheritance, forensic pathologist Bennet Omalu's contention that football players are at risk for chronic traumatic encephalopathy, and German cardiologist Andreas Roland

Grüntzig's proposal that blocked coronary arteries can be opened with balloon angioplasty were all initially scorned but are now widely accepted. When physicist Robert Goddard proposed in the 1920s that he could launch a rocket into space, he was excoriated by a *New York Times* editorial for not understanding that in the vacuum of space the exhaust of a rocket would have nothing to push at. That of course turned out to be wrong, and Goddard is now regarded as one of the pioneers of space travel.

There is no doubt that there have been many scientists who were at first mocked and later venerated. But I would venture to say that Jim Humble will not be one of those. It seems appropriate to end with Carl Sagan's famous comment: "The fact that some geniuses were laughed at does not imply that all who are laughed at are geniuses. They laughed at Columbus, they laughed at Fulton, they laughed at the Wright Brothers. But they also laughed at Bozo the Clown."

A MUSICAL MEMORY OF PLEXIGLAS

"Don't sit on the plexiglass toilet said the momma to her son!" That's the intriguing beginning to the lyrics of a song hidden in *The Serpent Is Rising* album by the American rock band Styx in 1973. When I say hidden, I mean that the song is not listed on the label and is only heard after playing "As Bad as This," the fourth song on side one of the album. The message of the song is very confusing when you consider verse two, which refers to a plexiglass toilet and a toilet lid landing on a boy's "banana."

One would think that if one fears injury to that particular part of the anatomy from an attack by a toilet lid, one would be motivated to sit down in order to avoid such a catastrophe. The scenario may sound odd, but in 2013 the *British Journal of Urology* published a paper entitled, "No small slam: increasing incidents of genitourinary injury from toilets and toilet seats." The researchers described some 9,000 cases of penile crush injury reported by U.S. emergency room physicians between 2002 and 2010, with most occurring in boys aged two to three years who were undergoing toilet training. Risk could be reduced, the authors concluded, by exchanging heavy toilet seats with slow-close toilet seat technology.

What has prompted this discussion of toilet seats and the rekindling of memories of the Styx album? The reference to plexiglass. It is hard to think of anything good about COVID-19, but it sure has spurred a booming market for plexiglass. Just about every store features plexiglass panels in some way to separate salespeople from customers. Given that we know how this virus is most likely to spread, basically by droplets expelled from the mouth when sneezing, coughing, or just talking, see-through barriers make a lot of sense. Especially when you don't have to worry about them breaking like glass. So, thank chemistry for poly(methyl methacrylate), the plastic commonly known as plexiglass.

As the name suggests, poly(methyl methacrylate) is made of repeating units of methyl methacrylate. German chemist Rudolph Fittig demonstrated the polymerization of methyl methacrylate in 1877, but it wasn't until the 1930s that the reaction was put to practical use simultaneously in England by Rowland Hill and John Crawford at Imperial Chemical Industries (ICI) and Otto Rohm in Germany. They were greatly helped by McGill University chemist William Chalmers's finding that methyl methacrylate could be easily produced from acetone and hydrogen cyanide, both of which were readily available. ICI trademarked the product as Perspex and Rohm named it Plexiglas, which remains a trade name. Plexiglass, with the double *s* ending, is a common name used for all acrylic plastics.

Plexiglas and Perspex played important roles in the Second World War with both the Allies and the Germans finding multiple uses for the novel plastic. Submarine periscopes, aircraft windshields, and gun turrets all benefited from the clear plastic that could be molded into desired shapes and was much tougher than glass, although it was not indestructible. Direct hits would shatter it and send tiny slivers flying everywhere. Some of these slivers, on occasion, lodged in the pilots' eyes. Usually, any foreign substance in the eye causes terrible irritation, but a British eye surgeon, Dr. Harold Ridley, noted that Spitfire pilots did not suffer the expected consequences. Somehow, their eyes tolerated this particular foreign material!

Ridley now had a vision. Maybe here was a way to solve the problem of cataracts, those opaque deposits that form in the lens of the eye as we age. The only method to treat cataracts at the time involved surgically removing the lens and fitting the patient with "Coke bottle" glasses, which would do the job that the natural lens had done. The widespread belief was that any kind of implanted lens was doomed to fail because the eye would reject it. But maybe it wouldn't reject poly(methyl methacrylate), Ridley thought.

In 1949, Ridley carried out his first successful Perspex implant. The plastic performed well, but the surgical techniques were not refined enough; the lens would often slip out of place and the trauma of the surgery led to all kinds of complications. Most of these problems were eventually solved by the Dutch ophthalmologist, Dr. Cornelius Binkhorst, making the implantation of acrylic lenses after cataract surgery routine.

Polyacrylics have since found their way into our lives in numerous other ways. Artificial teeth, eyeglasses, and the "glass" around hockey rinks and aquariums are just a few examples. One of the most popular tourist attractions in Japan is the Hipopo Papa Café, where you can sit on a toilet surrounded on three sides by a giant plexiglass aquarium. While answering nature's call, customers can enjoy the tropical fish swimming all around them.

I don't know if the seat itself is made of plexiglass, but it certainly could be. There are many commercially available plexiglass toilet seats, with some even featuring designs that give the appearance of an aquarium. More fun to use than ordinary white seats. And they can come equipped with slow-close lids so that little boys don't have to sit on the plexiglass toilet to avoid injury.

THERE IS MORE TO THE PLACENTA THAN THE MOTHER-EMBRYO CONNECTION

Vasily Zaytsev's vision was legendary. During the Battle of Stalingrad, the Soviet sniper eliminated over 200 enemy soldiers with a standard-issue rifle. His fame was such that his exploits were portrayed in a 2001 film, *Enemy at the Gates*, starring Jude Law as Zaytsev. During the war, the sniper was temporarily put out of commission when he suffered an eye injury from a mortar attack. Why temporarily? Because his vision was restored with an innovative treatment offered by Vladimir Filatov, a pioneer in the field of ophthalmology.

Filatov's major contribution was the corneal transplant. The cornea is the clear layer that covers the eye and together with the lens focuses light on the retina to produce an image that is then transformed into electrical impulses carried by the optic nerve to the brain, and we see. Injury to the cornea can affect vision, with the extent being determined by the degree of damage. Minor injuries heal by themselves, but major damage can require a corneal transplant.

The first such transplant using a cornea from a cadaver was attempted by Filatov in 1912 but was unsuccessful as the tissue quickly became opaque. Thinking that the freshness of the cornea to be transplanted is critical, Filatov experimented with preserving corneas from cadavers, eventually determining that at a sufficiently low temperature they could be kept viable for forty hours. By 1934, he was successfully transplanting corneas! Then he made an interesting discovery. When only a part of the cornea was replaced, the opacities on the remaining part cleared up! He concluded that the healthy cornea was releasing "biogenic substances" that helped heal diseased tissue.

Capitalizing on this observation, Filatov introduced "tissue therapy," a technique he claimed would improve various eye problems ranging from nearsightedness to retinal degeneration. All that was needed was the application of tissues that harbored these biogenic substances. He tried all sorts of biological materials, finding that placental extracts were the most suitable. Exactly what sort of treatment Filatov used on Zaytsev isn't clear, but it appears to have worked. What is clear is that for a brief time, thanks to widespread reporting of Filatov's ophthalmological successes, the treatment of eye diseases with topical placenta extracts achieved popularity in Europe. Before long, though, the claims of efficacy turned out to be shortsighted and ophthalmologists abandoned the technique.

The baton, however, was picked up by cosmetic chemists for whom the notion of the placenta containing some undefined "biogenic substances" struck a chord. After all, this is the organ that links mother and fetus, transferring nutrients, eliminating waste, and supplying various hormones and growth factors to the embryo. Could these substances be beneficial if applied to the skin? By 1950, the concept of placenta facials was introduced, with a number of European companies promoting creams formulated with purified human placenta extract and claiming efficacy based on cell proliferation studies in the laboratory. In 1958, Elixir Natale, containing "Placentine," a highly concentrated placenta extract, was introduced to the American market with claims of "providing rebirth of the skin and enabling it to remain in the bloom of babyhood." The cream would "overcome the effects of age on the skin and would impart a youthful elasticity which would banish the drying, faded look of age."

These claims did not sit well with the Food and Drug Administration (FDA) and shipments were seized under the 1938 Food, Drugs, and Cosmetics Act, which prohibited "false and misleading representations." The cream was renamed "Crème Paradox," and while it still contained Placentine, the extravagant promotions vanished. Whether the source of the placenta was human or animal was not identified, and since there was a lack of supporting documentation of benefits, the real paradox was why anyone should purchase the product.

Today, there are numerous cosmetic creams on the market that contain placenta components, and while the claims on the label tend to be rather innocuous, advertising on the web is highly imaginative, listing a tantalizing array of placental components that include stem cells, nucleic acids, peptides, superoxide dismutase, glutathione peroxidase, cytokines, collagen, elastin, vitamins, minerals, and growth factors. The product labels discuss the roles these play in biosynthesis, cell morphology, cell adhesion regulation, autocrine signaling, hydroxyl radical scavenging, cellular regeneration and, of course, "tissue rejuvenation."

All of this sounds enthralling to people for people who glance into the mirror and see crow's feet expanding into eagle claws. Unfortunately, there is no evidence for placenta extracts having a cosmetic benefit. Nevertheless, "placenta anti-wrinkle cream" on a label is allowed because any cream that prevents moisture loss from the skin can be labeled as "anti-wrinkle," since moisture plumps up the skin. For example, Rebirth, a cream that features "placenta anti-wrinkle" on the label, has mineral oil as the prime ingredient, followed by a host of other fatty substances that form a moisture barrier. Coming in twelfth on the list is hydrolyzed placenta protein. This is no different from any other type of protein, and any rejuvenating effect attributed to it is illusory. Still, there are many believers, especially celebrities who can afford to shell out hundreds of dollars for a sheep placenta facial that claims to tap into the power of stem cells. The claim that any type of stem cell topically applied will somehow produce youthful skin is pure folly. So is the supposed rejuvenating effect of the gold flakes that are mixed into the concoction.

SCIENCE FICTION, SCIENCE FACT

Long before COVID-19, and before he terrified movie audiences with carnivorous velociraptors in *Jurassic Park*, Michael Crichton frightened readers with the story of a murderous extraterrestrial microbe. In his 1969 novel *The Andromeda Strain*, a satellite returns to Earth contaminated with a microbe that quickly kills everyone in the vicinity of the town where it lands, save for the town drunk and a crying baby. Why did they survive? It all has to do with acidosis and alkalosis. Time for some simple physiology.

Blood is a very complex fluid containing numerous components, including various acids and bases. The acidity of any solution is described by means of the pH scale, on which 7 is neutral, below 7 signifies acidity, and above 7 indicates that the solution is basic or alkaline. The pH of the blood is critical in maintaining health, and calamities ensue when the pH ventures outside the normal range of 7.35–7.45. Acidosis can cause weakness, nausea, and confusion, while alkalosis is associated with irritability, muscle twitching, and convulsions. In the extreme, both conditions can be fatal.

The human body is a remarkable machine. Blood is actually a buffered solution, meaning it contains components that can react with excess base or excess acid to restore a normal pH. Carbon dioxide, which is produced by the metabolism of food, dissolves in blood to produce carbonic acid, which can neutralize any excess base. The bicarbonate ion, also present in blood, will promptly take care of any surplus acid.

The level of carbon dioxide in the blood can also be adjusted by our rate of respiration. If blood pH drops, which actually means that the blood has become more acidic, we breathe faster, exhale carbon dioxide, and thereby reduce the acidity. If the pH rises, respiration is inhibited, less carbon dioxide is exhaled, and again the pH returns to normal. Then there is a backup system. The kidneys can regulate blood pH by excreting or retaining acid.

This regulatory system is highly effective and can meet most challenges, but not all. Conditions that affect respiration, such as pneumonia, injury to the brain's respiratory center, or morphine overdose, can lead to respiratory acidosis. Uncontrolled diabetes, kidney failure, and starvation can all result in a decrease in blood pH, and cause metabolic acidosis. This can also happen from an overdose of alcohol or poisoning by antifreeze, since both of these substances are metabolized into acidic compounds.

Respiratory alkalosis is usually caused by excessive loss of carbon dioxide due to hyperventilation. That's why hysteria or panic attacks can have physical consequences and can be treated by breathing into a paper bag. The exhaled carbon dioxide is re-inhaled and reduces pH! Metabolic alkalosis can be caused by loss of hydrochloric acid from the stomach through persistent vomiting, kidney disease, or excessive intake of antacids.

These ideas were elegantly introduced to the general public in *The Andromeda Strain*. Michael Crichton was a physician with a sound scientific background, which he used to weave an ingenious plot. It seems that the invading microbe can only multiply effectively in human blood that has exactly the right pH. And guess what pH our invading virus prefers? That's right, 7.4, the pH of normal blood.

So, why did the baby survive? Because he was crying and screaming with all his might, and thereby exhaling a lot of carbon dioxide. This, as we have seen, raises the pH. The drunk, on the other hand, had run out of wine and was reduced to drinking Sterno! That's the little can of jellied fuel which warms many a buffet dish from underneath with a nearly invisible blue flame. The combustible ingredient is methanol, also known as wood alcohol. Metabolism in the liver converts methanol to formic acid. This acidifies the blood, alters the pH, and wreaks havoc. Drink enough methanol, and it will kill you. Our friendly drunk was lucky enough to indulge just enough to slightly alter the pH of his blood, which was enough to destroy the deadly microbe but not enough to kill him. A bit far-fetched, but at least rooted in real physiology.

However, Crichton travels deep into science fiction with his description of the microbe as having no DNA, RNA, or any proteins, and existing through its ability to transform energy into matter. Eventually, the Andromeda microbe mutates into a benign form, quite a feat without any genetic material. There is one interesting twist. Before becoming benign, the microbe mutates into a form that thrives on consuming plastic. We could use such a microbe today to rid the oceans of waste.

In 1971, *The Andromeda Strain* was made into a movie that was pretty close to the novel's storyline, except for the ending. In the film, clouds were seeded to precipitate rain and wash Andromeda into the alkaline water of the oceans. The pH of seawater is about 8.1 as a result of basic calcium carbonate leaching out of seashells, coral, and limestone deposits and would indeed be deadly to a microbe that can only survive in the 7.4 pH range. Enjoy the book, and watch the movie!

THE LEGEND OF JOHNNY APPLESEED

Those fluffy clouds you sometimes see up in the sky? They are really apple blossoms in Johnny Appleseed's heavenly orchard! That's the message delivered in Walt Disney's short 1948 animated musical *The Legend of Johnny Appleseed.* Young Johnny wants to go west with the pioneers but fears he is ill-equipped to survive in the wild. An angel appears and convinces him that all he needs is a bag of apple seeds, a Bible, and a pot to cook in that he can also use as a hat. Off he goes, spending a lifetime planting apple orchards until he takes his final rest under an apple tree. The angel appears once again telling him that his work on Earth is done, but there is a need for apple trees where they are headed. Johnny leaves behind his mortal body and, singing the catchy tune "The Lord is Good to Me," sets out to plant his heavenly orchard.

The Legend of Johnny Appleseed is rooted in reality. In the first half of the nineteenth century, John Chapman developed a passion for growing apple trees and traveled widely in the U.S., promoting the establishment of apple nurseries. He was a devoutly religious man and spread the Gospel along with his seedlings, as accurately portrayed by Disney. Chapman commonly wore simple clothes and often went barefoot, believing that the more he endured in this world, the greater would be his happiness in the hereafter.

Johnny is owed a debt of gratitude for rehabilitating the image of the apple, a fruit that historically did not have a stellar reputation thanks to its identification as the forbidden fruit of the Bible. Actually, the Bible does not name the fruit with which Eve tempted Adam, but when artists in the Middle Ages began to paint biblical scenes, they portrayed the apple as the fruit of the tree of knowledge, possibly due to a misunderstanding in a Latin version of the Bible in which "good and evil" is noted as "boni et mali." In Latin, apple is *malus*, with its plural being *mali*. The image of the apple also suffered from the folktale that the Adam's apple in the male throat was caused by the forbidden fruit sticking in Adam's throat. So, the apple was in need of some re-imaging, and John Chapman planted the seeds for that.

Why Chapman was so enamored of apples isn't clear, but likely was due to the ease with which apple trees grew in the northern U.S. There is no evidence that he associated the fruit with health; that notion was first formulated in a Welsh proverb tracing back to 1866: "Eat an apple on going to bed, and you'll keep the doctor from earning his bread." This would eventually be shortened to the more common "An apple a day keeps the doctor away." Does it really?

Believe it or not, this has been examined. Dartmouth researchers plucked relevant data from the U.S. National Health and Nutrition Examination Survey (NHANES), in which some eight thousand subjects filled out questionnaires about their daily food consumption, including fruits. About 750 people reported that they regularly ate an apple a day. When they were questioned about the number of visits to a doctor during the previous year, there was no difference between the apple eaters and the non-apple-eaters. On the other hand, the apple eaters were marginally more likely to avoid prescription drugs. So, while an apple a day may not keep the doctor away, it may keep the pharmacist at bay. Calculations showed that if everyone did eat an apple a day, it would lower prescription costs in the U.S. by about $47 billion. There is a confounding factor here, though. Apple eaters were also less likely to smoke.

If Johnny Appleseed happens to glance down from his heavenly orchard, he might be surprised by some of the concerns being raised about eating apples. There is the issue of pesticide residues, with activists pointing out that there are over thirty pesticides registered for use on apples. Of course, only a few of these may actually be used by a grower, and analyses of residues show they are well below the acceptable daily intake levels. Recently another concern has arisen. Researchers have found that apples contain more microplastic particles than other fruits. It is somewhat of a mystery why apples accumulate these particles, which originate from the breakdown of plastics in the environment, but they are probably taken up from water traveling up the root system. Whether the 200,000 or so microparticles per apple detected through electron microscopy have any health consequence is unknown, but it certainly strengthens the case for reducing plastics in the environment.

There is also some positive information emerging about apple consumption. Apples, particularly around the core, contain a bevy of diverse, harmless bacteria that can help populate the gut and starve out disease-causing germs. An apple has about 100 million bacterial cells, 90 percent of which are in the core, suggesting that we have been eating apples the wrong way. We should start at the bottom and eat up through the core! And don't worry that the seeds contain cyanide. You would have to eat a cupful of them to cause harm and they would have to be chewed to release cyanide from amygdalin, the cyanide-containing compound.

Today there are some 7,500 varieties of apples in the world created through grafting. Celebrated food writer Michael Pollan argues that Johnny Appleseed did not carry out grafting and only planted seeds. Those available at the time would have resulted in apples that were too sour to eat but would have made perfect cider! Maybe that is why Johnny achieved such fame among the pioneers!

TREADING GINGERLY ON GINGER

The first time I got interested in ginger was on a trip to New York City in 2013. I was keen on touring the New York Hall of Science, a museum that had its origins in the 1964 New York World's Fair, which I also had attended. Back then I was a big fan of the space program and was very impressed by the rockets on display, especially the Mercury capsule that had carried astronaut Scott Carpenter into Earth orbit in 1962. I was anxious to see the Rocket Park again and was not disappointed. It had been expanded with pristine versions of the Titan II and Atlas rockets that had carried Gemini and Mercury capsules into space. There were also various exhibits about molecules, microbes, and inventions. I was especially drawn to Gingerbread Lane, a gingerbread village created by chef Jon Lovitch; according to the description, it weighed 1.5 tons and covered 300 square feet. It was certified by Guinness World Records as the largest ever gingerbread village!

Gingerbread can be traced back to the eleventh century and has been a holiday favorite ever since, partly because besides imparting flavor, ginger also acts as a preservative. The tradition of making gingerbread houses seems to be connected to Hansel and Gretel, the Brothers Grimm tale in which two children abandoned in a forest come upon an edible house decorated with candy. German bakers took to baking gingerbread houses, capitalizing on the story's popularity.

What we refer to as ginger is the underground stem of a plant native to Asia. It has a long history of use not only as a spice but also as a folk medicine, mostly for digestive problems and nausea. Like any plant, ginger is composed of a large number of compounds, and ginger's pungency and anti-nausea effects are due to gingerol. When ginger is heated, gingerol is converted into zingerone, as well as into a class of compounds known as the shogaols, all of which contribute to flavor and pungency.

Several lab studies have examined ginger's potential for biological activity and managed to find some antibacterial and anticancer effects, but this is not surprising. There are numerous compounds that have such activity in vitro —in the lab or outside or living organisms — but rarely translate to any clinical benefit. The most significant effect that has been demonstrated in people, although not in all trials, is some efficacy in treating nausea during pregnancy and chemotherapy. Ginger has a tradition as a stomach settler, and shogaols specifically are claimed to have an anti-coughing effect, but the evidence is anecdotal.

Ginger also has some irritant properties, something that once found an unusual application. In the sixteenth and seventeenth centuries, ginger was used to make a horse carry its tail high and move in a livelier fashion. How? By inserting into the horse's rectum a piece of ginger whittled into a suppository-like shape! Horse traders would do this to make an older horse behave like a younger one or to temporarily liven up a sick animal. The process was known as "feaguing." Apparently, although illegal, ginger-based creams are applied to the rectum of show horses even today. If a cream is detected the horse is disqualified. In days of yore, there was something else used on horses in a fashion similar to a piece of ginger. A live eel was a physical irritant that made the horse step more sprightly.

These days, it seems it is not only horse breeders who are into gingering. Now also known as "figging," some people engage in the practice because they find the burning sensation produced by ginger, albeit painful, also erotic. Each to his own.

Believe it or not, "eeling" is also a real phenomenon. There are several cases recorded in the medical literature documenting the removal of eels from the rear portals. In one instance, emergency room physicians in a New Zealand hospital had to extricate a wriggling creature from a patient. Respecting the patient's privacy, details about how the eel ended up in the man's nether region were not revealed, but a spokesperson confirmed that "the eel was about the size of a decent sprig of asparagus and the incident is the talk of the place." I bet it was.

In another episode in China, a factory worker was rushed to a hospital after complaining of a terrible stomachache. Doctors were shocked to find a half-meter-long expired Asian swamp eel in his intestines, along with fecal matter. At first the man claimed the eel entered by mistake, later admitting that he had tried to cure his constipation in this unique fashion.

A more traditional way to address digestive problems is by drinking ginger ale. Many people apparently purchase the beverage with the belief that the ginger it contains has various health attributes and "settles the stomach." When some consumers discovered that the "natural flavor" responsible for the taste only contains about two parts per million of actual ginger extract, which is less than what humans can detect, and is way too little to have any health benefit, they decided to launch a number of class action lawsuits. They argued that the label stating "made with real ginger" was misleading for people who believed the drink could help with their digestive issues. In a recent Canadian case, Canada Dry settled for $200,000 with a man who claimed the company had engaged in false advertising. One wonders if Canada Dry executives tried to settle their stomach with ginger ale after that settlement.

GORILLA GLUE MAKES FOR A BAD HAIR DAY

Substituting Gorilla Glue for hair spray is a decidedly bad idea. Just ask Tessica Brown, the Louisiana woman who made this mistake and ended up gluing her hair strands together and cementing them to her scalp. Talk about a bad hair day!

The surface of any material, no matter how smooth it may seem, is pockmarked with tiny pores. Adhesives are chemicals that are fluid enough to fill these pores on the surfaces that are to be glued together and then set into a hard matrix. Gorilla Glue is a liquid mixture of a diisocyanate and a polyol, chemicals that react to form a rigid polyurethane polymer when activated by moisture. Once the polyurethane fills the pores, it expands into a foam as carbon dioxide is released during the polymerization reaction. The pressure exerted against the walls of the pores and the so-called van der Waals attraction between the polyurethane and the surface ensure a tight bond between the glued materials. It is like using thousands of microscopic screws! In the case of the unfortunate Ms. Brown, the polyurethane formed a tough layer of glue and matted hair that painfully tugged at her scalp.

As I am fond of saying, there are no good or bad chemicals, just proper and improper ways to use them. Indeed, polyurethane is one of the most useful synthetic substances in existence, with applications that extend far beyond adhesives. For me, this polymer has very special significance because it was one of the first chemicals that really caught my attention. That memorable event occurred at the New York World's Fair in 1965, where the DuPont pavilion featured a live musical extravaganza with dancers singing about "The Happy Plastics Family." The entertainment began with, "E.I. du Pont de Nemours & Company Wishes you the most entrancing hour that there could be / Where imagination is fancy-free / In the 'Wonderful World of Chemistry.'" Kicking up their heels, the dancing molecules continued: "With Antron and Nylon and Lycra and Orlon and Dacron / The world's a better place / You know we all have a smile on / That started with Nylon / And stretches across each happy face."

That was all quite enchanting, but the stanza that really captured my imagination went like this: "We're gonna have shoes like we never had shoes before / We're spreading the news, we'll be walkin' on clouds galore / So let it rain or let it shine / It's very plain the weather's fine / With shoes for showin' the blues the door." Those shoes were made of DuPont's novel material, Corfam.

The company had become famous in the 1930s with Wallace Carothers's discovery of what would become DuPont's flagship product, nylon. Of course, this (as well as polyesters that Carothers had also developed) was protected by patents that other companies tried to circumvent. In Germany, Otto Bayer, working for the IG Farben Company, was trying to do just that when he discovered that polyisocyanates react with polyols to form polyurethane. The Germans used this chemistry to produce coatings for aircraft during the Second World War, and the material also proved to be useful for insulating beer barrels!

By the 1950s, DuPont had concluded that "leather seemed ripe for substitution," and polyurethane was just the material to introduce revolutionary footwear. To overcome the problem of sweaty feet, fake leather, which had been trade named Corfam, was pierced with millions of microscopic holes to make for a "poromeric" material, a term coined from "polymer" and "porous." The shoes were water-repellent, were supposed to never need shining, and would last longer than leather. That was one of the problems. People did not want everlasting shoes, they wanted to embrace new styles. There were also complaints that the shoes would never break in, although policemen who were secretly enlisted to try them were satisfied. Neither were there complaints from the dancers in the show, who seemed to frolic happily in their Corfam shoes. The public though did not take to the newfangled footwear, and in 1970, DuPont divested itself from Corfam, making for a business debacle on par with Ford's Edsel and Coca-Cola's New Coke. Today, some companies still produce polyurethane shoes, mostly for the military where they make for an easy "spit and polish" shine.

Corfam turned out to be polyurethane's sole failure. Today, this multi-functional polymer is used in a myriad of items that include automobile seats, varnishes, soccer balls, sponges, condoms, skateboard and roller coaster wheels, tires, spandex fiber, surfboards, car bumpers, insulating materials, mattresses, pillows, catheters, shoe soles, and of course, as Tessica Brown discovered, extremely strong adhesives.

The unlucky woman's problem was finally solved by Beverly Hills plastic surgeon Dr. Michael Obeng, who cleverly made use of his undergraduate degree in chemistry. Dr. Obeng heard about the lady's catastrophic experience and came to her rescue with what the press reported as surgery, although there was actually no surgery involved. Dr. Obeng experimented with human hair on a mannequin to find solvents that would dissolve polyurethane, and while no exact description has been given, it seems that a combo of acetone, petroleum distillate, and medical adhesive remover did the trick. It took four hours of careful rinsing with the solvent mixture, but eventually Ms. Brown was able to happily run her fingers through her hair. Puts a whole new meaning on "splitting hairs."

BRUSHING UP ON HAIR DYES

"Now we will find out what people's natural hair color is." That joke scooted around the blogosphere after hair salons were forced to close down in response to COVID-19. However, not everyone was dismayed by the reduction in the use of hair dyes. Critics of the cosmetic industry pointed out various concerns that had been raised over the years about para-phenylenediamine (PPD), a common ingredient in permanent hair dye formulation. They welcomed a decline in exposure to this chemical.

Para-phenylenediamine first came to the public's attention in 1933 at the Chicago World's Fair, where a dramatic exhibit by the Food and Drug Administration caught the eye of visitors. On display were pictures of women who had suffered eye damage, and in one severe case, blindness, after having used "Lash Lure," an eyelash and eyebrow dye. The exhibit was stimulated by a couple of recent reports in the *Journal of the American Medical Association* (*JAMA*) that described severe eye injuries after the use of products that contained PPD. At the time, there were no regulations pertaining to cosmetics, and manufacturers had no obligation to test their products for safety. The FDA had long been clamoring for a change to the Food and Drug Act of 1906, the only relevant law in existence, but it had little teeth and ignored cosmetics. The World's Fair exhibit was designed to bring to the public's attention the dangerous consequences of a lack of proper regulations.

The display received a boost in publicity when First Lady Eleanor Roosevelt visited and exclaimed, "I cannot bear to look at them" on seeing the pictures. A reporter quickly dubbed the exhibit "The American Chamber of Horrors." The public got further wind of what was going on when Paramount released a newsreel including footage of the corneal damage suffered by women using the "eyelash beautifier." The film did not identify the product as Lash Lure, which led to Maybelline, manufacturer of a competing product, to protest vigorously to Paramount. They complained that although their product contained no PPD, their reputation and sales had been damaged because they had used "eyelash beautifier" in their advertising. Maybelline actually used coal dust mixed with Vaseline, an idea that came from Mabel, the sister of the company's founder.

In 1934, another paper appeared in *JAMA*, this time documenting the death of a woman after her eyebrows had been completely plucked and then replaced by the dye. It seemed that the PPD had caused a severe allergic reaction, irritating the skin and allowing bacteria to, enter, causing sepsis. In those pre-antibiotic days, such infections were often fatal. President Roosevelt was a strong proponent of public health and in 1938 managed to have Congress pass the Food, Drug and Cosmetic Act, which prohibited the use of PPD in eye cosmetics but allowed its use in hair dyes, where it can be found to this day.

Para-phenylenediamine was first synthesized in 1854 from the coal tar chemical aniline by German chemist August Wilhelm von Hofmann. It is unclear to what extent von Hofmann investigated the properties of the chemical, but we know that in 1883, the French company P. Monnet et Cie was granted a patent for its use as a dye. When PPD is exposed to oxygen in air, it forms a number of compounds ranging in color from dark brown to black. Since hydrogen peroxide releases oxygen, combining it with PPD speeds up the development of color. This proved to be valuable in the fur trade, and by the late 1800s, PPD was also being used to dye hair. Results were not always satisfactory since the color washed out too easily.

Researchers eventually found that when PPD is mixed both with hydrogen peroxide and a coupler before being applied to the hair, a reaction takes place that produces a dye, but only once the chemicals have been absorbed into the hair shaft. The colored molecules formed are too large to diffuse out of the hair, hence the term "permanent." Temporary or semi-permanent dyes do not contain PPD.

Para-phenylenediamine is a documented allergen, which of course is an issue, but there is also concern that it may be a carcinogen. The Sister Study, published in 2019, surveyed a large cohort of American women who had a sister with breast cancer but were cancer-free themselves. It found an increased risk of cancer with permanent dye use, especially in Black women. A previous study by the Rutgers School of Public Health in 2017 looked at some 4,000 women, half of whom had breast cancer, and came to a similar conclusion. Of course, such associations cannot prove cause and effect due to confounding factors such as diet and the use of other personal care products, such as hair-relaxers, that can contain formaldehyde.

Although there is no proven link between breast cancer and permanent hair dyes, researchers are looking at alternative formulations that eliminate PPD. Polydopamine can mimic natural melanin in hair, and anthocyanins derived from black currants also have potential. In any case, the risk, if any, from permanent hair dyes is small, and there is also something to be said for being happy about what you see in the mirror.

JEAN HARLOW'S HAIR

Howard Hughes, who would go on to become a wealthy business magnate, engineer, pilot, philanthropist, and famous sufferer of obsessive-compulsive disorder, started out as a film producer. That's when he came across Jean Harlow, a young ash-blonde actress who had the makings of a star. Mary Pickford had gained spectacular fame as "America's Sweetheart" (she was actually Canadian) and Hughes thought Harlow could follow in her footsteps if she were properly promoted. That challenge was met by his publicity director, who came up with the moniker "Platinum Blonde."

The metal platinum derives its name from the Spanish *platino* for "little silver" thanks to its almost white luster. It also has an aura of wealth due to its rarity and seemed just the right description for a potentially shining star. But there was just one problem. Harlow was indeed blonde, but not uniquely blonde. She wasn't "platinum" enough. So the hairdressers went to work and after some experimentation were able to rid her hair of all the natural melanin pigment, resulting in a silvery-white color. Now the publicity team swung into action, offering a prize of $10,000 to any hairdresser who could match Harlow's shade. None could, which was probably a good thing, otherwise many women would have suffered the same fate as Harlow. Hair that was the texture of straw and prone to falling out. As was revealed decades later by her personal hairdresser, the sex bomb's famous platinum shade was achieved with a mix of hydrogen peroxide, sodium hypochlorite bleach, ammonia, and Lux flakes. That mix may have done more than just slowly destroy her hair, it may even have affected her health.

Jean Harlow died tragically at the age of twenty-six. She had suddenly become ill during the filming of *Saratoga* and had to be hospitalized. Three days later, she lapsed into a coma and died. The cause of death was kidney failure, possibly the result of a number of illnesses from which the actress had suffered throughout her life. She had multiple bouts of influenza, contracted scarlet fever when she was fifteen, and had a severe infection after removal of her wisdom teeth. But there is also a theory that her frequent hair treatments may have contributed to her demise.

When ammonia is mixed with hypochlorite bleach, there are some nasty compounds that can form, with chloramine, hydrazine, and hydrochloric acid among them. Chronic inhalation of these and absorption through the scalp can indeed put a burden on the kidneys. Whether this contributed to Harlow's health problems we will never know, but we do know that mixing ammonia with bleach is a bad idea. Both of these can be found separately in cleaning agents and should never be combined. Indeed, hypochlorite bleach should never be combined with anything. Mixing it with any acid, whether that is vinegar or rust remover, produces potentially lethal chlorine gas. Combining bleach with hydrogen peroxide yields oxygen gas, which is not a problem unless the mix is stored in a closed container. Then the pressure can build up and cause the container to burst.

While mixing hydrogen peroxide with other chemicals can cause problems for people, not so for the aptly named bombardier beetle, a little creature that can dispense chemical bombs to ward off its predators. When in danger of being attacked, the beetle produces a smelly, burning-hot toxic liquid that it sprays at the enemy in bursts from a turret-like appendage on its abdomen. There is some very interesting chemistry here. In one compartment the beetle stores a mix of hydrogen peroxide and hydroquinone, both metabolic products of its diet. Another vesicle contains enzymes that can quickly break hydrogen peroxide down into water and oxygen. When attacked, abdominal contractions mix the chemicals and the oxygen produced converts the hydroquinone into toxic quinone. The reaction is highly exothermic, and any predator that gets sprayed with this hot mix learns to leave the bombardier beetle alone. This chemistry can be life-saving for the beetle even if a predator has not been deterred. Frogs have been known to regurgitate beetles they've swallowed after the beetle exploded its chemical bombs in the frog's stomach.

Humans have also made use of hydrogen peroxide's ability to oxidize other chemicals. During the Second World War, the Germans developed the Komet, a rocket plane capable of achieving speeds of over 500 miles per hour, unmatched by any other aircraft. Concentrated hydrogen peroxide and hydrazine were stored in separate containers aboard the aircraft, and when combined they generated steam and nitrogen gas at high pressure. As these gases escaped from the combustion chamber, they propelled the vehicle forward according to Newton's third law, which states that there is an action for every reaction. Although the Komet could readily pick off slower flying Allied aircraft, it never met with great success because mixing the peroxide with the hydrazine often led to the Komet itself exploding due to the instant buildup of high pressure.

Incidentally, Howard Hughes also discovered a young brunette who would go on to achieve fame after becoming a blonde using the same chemistry as Jean Harlow. Marilyn Monroe in her autobiography opined that "in Hollywood a girl's virtue is much less important than her hairdo."

OH, THE DREAD OF RED M&MS

It is always fun to search for flubs in movies. Keen eyes have alerted us to a Roman soldier wearing a watch in *Spartacus*, a gas cylinder visible in an overturned chariot in *Gladiator*, a car in the background in *Braveheart*, bullet holes present in the wall before shots are fired in *Pulp Fiction*, and a coffee cup making an appearance in *Game of Thrones*. More recently, in an episode of the hit science fiction series *Stranger Things*, set in the early 1980s, young Mike Wheeler pours out some M&Ms with a red candy being clearly visible. The problem is that were no red M&Ms at the time; the color had been eliminated in 1976. That was because the U.S. Food and Drug Administration had banned the use of Red Dye No. 2, also known as amaranth, as a result of safety concerns that had been raised. (Amaranth dye is named for, but not related to, the amaranth plant.) Although M&Ms had never used this dye, the massive publicity that had been given to the banning of amaranth raised concern at Mars and Murrie, the makers of the candy-coated chocolates. They were concerned people would avoid all red colorants, including Red No. 3 (or erythrosine), the color the company used, so they decided the world could do without red M&Ms.

The banning of Red No. 2 was controversial for several reasons. It had been widely used since the passage of the Pure Food and Drug

Act of 1906 in the U.S. After examining some seven hundred or so synthetic colors in use at the time, allowable ones were reduced to seven, including amaranth. There were no concerns raised about this dye until the early 1970s, when a couple of Soviet studies claimed that the dye caused cancer and reproductive problems in rodents. The U.S. at the time was suspicious of any information that emerged from the Soviet Union. After all, what did the Soviets know about red? American scientists were quick to criticize the studies for poor methodology, but nevertheless the FDA decided to carry out its own studies. Most did not find any problem; however, a few did show some toxic effects, albeit at unrealistically high doses. Given that other red dyes were available, and that some studies had raised the red flag, the FDA decided to ban amaranth. The industry mostly switched to Red No. 40, also called "allura red," which curiously was banned in many European countries. Canada allows both Red No. 40 and Red No. 2, but the amounts allowed and the foods in which they can be used are carefully defined.

The first law that concerned food colorants seems to be one passed in Germany in 1531, and it was pretty harsh. Anyone caught counterfeiting saffron, used at the time as a yellow dye, was to be burned. By the late 1800s, Germany and England had passed regulations against the use of red lead oxide and red mercury sulfide to color confections, and against green copper arsenite to dye pickles or used tea leaves for resale. Synthetic dyes were introduced after William Henry Perkin's accidental discovery of mauveine in 1856. These "coal tar" dyes, so-called because the starting materials for their synthesis were isolated from the residues of burned coal, proliferated widely until the U.S. focused attention on them with the Pure Food and Drug Act of 1906.

Food dyes flew pretty well under the radar until 1950, when children across the U.S. came down with diarrhea and rashes after eating candy and popcorn balls tinted with the FDA-approved Orange Dye No. 1, one of the seven colors that had been approved in 1906. The

FDA launched an investigation and found that manufacturers were using huge amounts of the dye, doses that approached those known to cause severe toxic reactions in rodents. In 1956, Orange No. 1 was removed from the market. Things were quiet until the Soviet studies shook the U.S. and spawned the furor that resulted in the ban of Red No. 2 and the elimination of red M&Ms.

In 1987, the red M&Ms would make a triumphant return thanks to the antics of Paul Hethmon, then a student at the University of Tennessee. More or less as a joke, in 1982 he had founded the Society for the Restoration and Preservation of Red M&Ms, whose members were urged to bombard Mars and Murrie with letters. Eventually, the company relented and reintroduced the red candies, this time dyed with Red No. 40, except in Europe where this dye is not allowed. There the candies are dyed with carmine, extracted from the crushed bodies of female cochineal bugs that live on prickly pear cacti.

While it is extremely unlikely that Red No. 40 in M&Ms presents any risk for humans, the same may not be true for hummingbirds. These fascinating little creatures are known to be attracted to red colors, which explains why hummingbird feeders are colored red, and nectar sold to be used in feeders is red, usually colored with Red No. 40. This has resulted in many online discussions of this practice, with nectar sellers offering rewards to anyone who can provide a reference to a study that documents harm, and hummingbird lovers pointing out that while there may not be any proper studies, the doses the birds ingest per bodyweight are much greater than what causes toxicity in rodents. In any case, while it is clear that a red feeder attracts the birds, there no evidence that providing red nectar increases the attraction.

Incidentally, there is no truth to the rumor that the red M&Ms disappeared because the dye is an aphrodisiac and workers were stealing them off the assembly line. But I've heard whispers about the blue ones . . .

MYSTERIES, MAGIC, AND THYROID DISEASE

Dorothy Sayers is one of my favorite mystery writers mainly because she often weaves chemistry into her stories. The plot of *Strong Poison* revolves around arsenic poisoning, and in *The Documents in the Case,* a criminal is caught based upon a subtle difference between muscarine found in poison mushrooms and its synthetic analog. In *The Incredible Elopement of Lord Peter Wimsey*, the crime is not about a poison being administered, but rather about a necessary medication being withdrawn. That medication is thyroxin, needed to replace a deficiency caused by an improperly functioning thyroid gland.

Lord Peter Wimsey is Sayers's version of a detective in the mold of Sherlock Holmes or Hercule Poirot. He is a gentleman who, as he says, likes "to look into queer things." In this story, he encounters Langley, a professor of ethnology who has just returned from a Basque village where he accidentally chanced upon a couple he had been friendly with back in England. The lady, Alice Wetherall, was in a terrible state, "her face white and puffy, the eyes vacant, the mouth drooling and a dry fringe of rusty hair clinging to a half-bald scalp." Her skin was clammy to the touch and her behavior smacked of what in those days was called "imbecility." The story among the locals was that she had been bewitched. Wimsey didn't buy that, and furthermore, thought he had recognized the symptoms as someone suffering from myxedema, a case of severely advanced hypothyroidism. But the lady's husband was a physician — surely he would have recognized the symptoms!

Lord Peter thought the case was worth investigating. Some snooping revealed that the doctor had become insanely jealous of Langley, thinking that his wife had shown some interest in the gentleman. To put an end to the nonexistent romance, he spirited her away from England to the Basque village and punished her by taking away the thyroxin wafers she had been prescribed to combat her congenital hypothyroidism. As Wimsey would later explain to Langley, "Alice Wetherall is one of those unfortunate people who suffer from congenital thyroid deficiency. You know the thyroid gland in your throat — the one that stokes the engine and keeps the old brain going. In some people the thing doesn't work properly, and they turn out cretinous imbeciles. Their bodies don't grow, and their minds don't work. But feed 'em the stuff, and they come absolutely all right — cheery and handsome and intelligent and lively as crickets. Only, don't you see, you have to keep feeding it to 'em, otherwise they just go back to an imbecile condition."

Once Wimsey had gotten a grasp of the situation, he came up with a scheme to rescue the lady. The chance came when her husband was away in America, having left a housekeeper in charge of his wife with explicit instructions that she was to see no one. Wimsey traveled to the village pretending to be a wizard and amazed the folks with some magic tricks. Having witnessed the wizard's miracles, the villagers had no problem believing his claim that he was also able to counter spells. The credulous housekeeper was taken in and agreed to give her mistress wafers the wizard claimed would break the spell. And they did! After this, he had no trouble convincing the housekeeper to bring the lady to him so that he could permanently remove the spell. This was to be done in a special cabinet into which the wizard guided the bewitched victim. There was a puff of smoke, and when the housekeeper opened the cabinet's door, both her mistress and the wizard had vanished! The doctor on his return from America found nothing but an empty house and a legend. Lord Peter had orchestrated a magical elopement!

The science in Dorothy Sayers's story is sound. Although the history of thyroid disease goes all the way back to the Roman natural philosopher Pliny, who recommended treating a swollen thyroid, or goiter, with burnt seaweed, significant developments occurred in the early twentieth century that piqued Sayers's interest. In 1914, E.C. Kendall in the U.S. isolated the thyroid hormone thyroxin from 6,500 pounds of hog thyroids, and within three years it was introduced into commercial production. A molecule of thyroxin contains four atoms of iodine, which explains why burnt seaweed worked as a treatment for goiter. Seaweed contains iodine, and goiter is caused by iodine deficiency in the diet. As the thyroid struggles to extract iodine from the bloodstream, it enlarges. Iodine is present in soil near the world's seas and oceans but not inland, which is why goiter used to be endemic in mountainous areas like the Alps and the American Midwest. That problem was solved with the addition of iodine to salt. Today there is no need to extract thyroxin from animal thyroids; it can be made synthetically in the lab and is familiar to many as Synthroid.

I liked this story because of its scientific and magic connections. When discussing his conjuring abilities, Wimsey makes a reference to Jasper Maskelyne, a famous British stage magician of the times, and to David Devant, who had devised the cabinet used in the wizard's vanishing illusion. Indeed, Devant, considered to be the dean of British magicians, designed all sorts of illusions, including various contraptions used for disappearances. One of his signature tricks was the Obliging Tea Kettle, from which he poured any drink the audience asked for, be it water, beer, wine, or tea.

LICE, TYPHUS, AND NAPOLEON

Scottish physician James Lind is famous for finding the link between diet and scurvy. In his 1753 publication *Treatise of the Scurvy*, he described a series of systematic experiments he carried out while working as a naval surgeon on HMS Salisbury. Six pairs of seamen suffering from scurvy were given a remedy that had been proposed over the years by various medical authorities. Vinegar, garlic, cider, horseradish, mustard, and dilute sulfuric acid proved useless, but the men who were given oranges and lemons quickly recovered.

Not as well-known as his work on scurvy is Lind's contribution to the treatment of another scourge that plagued sailors and soldiers. Typhus presents with a fever, a body rash, and sensitivity to light before progressing to delirium and possibly death. Lind had noted the disease was much less likely to occur in hospital wards where patients were bathed, given laundered clothes, and slept in clean beds. This led to a recommendation that sailors be regularly scrubbed and issued clean clothes and bedding. As a result, British sailors were spared of "ship's fever," giving them a clear advantage in naval battles over the French, whose vessels were riddled with typhus. "War fever" was another description of the disease, one to which Napoleon could relate. His army of close to half a million soldiers had to retreat from Russia in 1812 because of an epidemic of typhus. Only about 6,000 managed to return to France.

Although the ancient Greeks described symptoms consistent with typhus, it was in the sixteenth century that the disease, likely brought back by the Crusaders from the Middle East, was documented in Europe. An epidemic in England that came to be referred to as the Black Assizes was the result of the disease being introduced into the courtroom by prisoners who had been incarcerated in filthy cells. Spectators, jurors, and judges contracted "jail fever," another name that would enter the lexicon for typhus.

Although it was clear that crowded, unhygienic conditions gave rise to the disease, the actual cause was unknown until 1909, when French bacteriologist Charles Nicolle identified body lice as the vectors of typhus. He had noted that infected patients, as well as their clothes, were able to infect others. An examination of the clothes revealed that they were infested with lice. Nicolle managed to show that after being given a hot bath and a change of clothes, typhus patients would no longer be able to spread the disease. This suggested an experiment in which he infected a chimpanzee with blood from a typhus patient, retrieved some lice from the animal's body, and placed them on a second chimp. Within ten days the second chimp came down with the disease.

Just a year after Nicolle had found the role lice played in the transmission of typhus, American pathologist Howard Taylor Ricketts determined that the infective agent was actually a bacterium that first infected lice and then their human hosts. It was eventually named *Rickettsia prowazekii* after him and Polish scientist J.M. von Prowazek, who had also investigated typhus. While both men achieved well-deserved fame, they paid for it with their lives. Both died of typhus contracted during their research.

Nicolle tried to develop a vaccine using crushed lice and serum from patients who had recovered from typhus, an early attempt at the use of convalescent plasma. He tried it on himself and on a few children, but the vaccine did not prove to be effective. In 1937, Herald R. Cox of U.S. Public Health Service did introduce a vaccine made from cultured bacteria that was widely administered to all Allied military personnel. While it did not prevent the disease, it did reduce its severity if contracted. A far more significant measure was eliminating lice with DDT powder blown under clothing. Incredibly, Allied forces in the Second World War had no deaths due to typhus, and DDT also stopped a terrible epidemic in newly liberated Naples during the winter of 1943–44.

Nazi concentration camps were a hotbed for typhus. Hundreds of thousands died from the disease, including Anne Frank, who succumbed in the Bergen-Belsen camp in 1945. When the camps were liberated, survivors were doused with DDT to prevent further spread of typhus. During the war, the Nazis had used the fear of typhus to relocate Jews in ghettos by promoting the perverse ideology that they were carriers of the disease. Confinement, such as 400,000 people packed into 1.3 square miles in the Warsaw Ghetto under starvation conditions, did cause an epidemic of typhus. But instead of accelerating as winter approached, the epidemic suddenly declined. This seems to have been due to doctors in the community recommending social distancing, isolation of infected people, monitoring for lice, and washing, at least as much as was possible under the terrible conditions the Nazis had set up to promote their propaganda of Jews being vectors for disease. Ending the outbreak thwarted that plan, but the story does not have a happy ending. The Nazis eventually used the fear of typhus to eliminate the ghetto and deport the inhabitants to concentration camps.

While DDT was very effective, it was eventually discontinued because of environmental concerns, lice developing resistance to the insecticide, and possible risk to humans. The typhus problem was finally solved in 1947 with the introduction of broad-spectrum antibiotics capable of killing the *Rickettsia* bacteria.

NAVALNY AND NOVICHOK

Alexei Navalny took a few sips of tea at the airport in Tomsk, Siberia, before boarding a plane bound for Moscow. On the flight he became so ill that the plane had to make an emergency landing in Omsk, where the doctors suspected but were unable to confirm some sort of drug overdose. Since Navalny was an activist who had been investigating government corruption in Russia and had previously claimed to have been sprayed by some sort of toxic chemical by assailants, a German humanitarian organization, Cinema for Peace, took an interest in the case and chartered a plane to bring Navalny to Berlin. Here doctors concluded that the symptoms, namely vomiting, sweating, respiratory distress, pinpoint pupils, foaming at the mouth, and a slower than normal heart rate, were consistent with the inhibition of cholinesterase, an enzyme that normally degrades the neurotransmitter acetylcholine. This suggested possible exposure to a nerve agent, a chemical that interferes with the transmission of information from one nerve cell to another.

Such transmission involves the release of chemicals called neurotransmitters from a nerve ending, followed by the migration of this substance across the tiny gap separating nerve cells, known as the synapse. The neurotransmitter then stimulates an adjacent cell by fitting into a receptor site on its surface, very much as a key fits into a lock.

Acetylcholine, the first neurotransmitter ever discovered (1921), stimulates muscle contraction, increases bodily secretions, pinpoints pupils, and slows heart rate. Once acetylcholine has carried out its job of triggering a reaction in an adjacent cell, it is decomposed by an enzyme present in the synapse. Overstimulation is therefore prevented. It is this enzyme, acetylcholinesterase, that is inactivated by nerve agents. Unless this activity is restored, overstimulation by acetylcholine will lead to convulsions, paralysis, and respiratory failure.

Blood tests carried out by a special military lab in Germany confirmed the presence of a nerve agent in Navalny's system. Furthermore, the chemical belonged to a series developed by the Soviets in the 1970s named Novichok, which in Russian means "newcomer." These newcomers were more potent than existing nerve agents and could be disseminated as powders and liquids. The Russians had used one of these chemicals before in the celebrated case that has come to be known as the Salisbury Poisonings in the U.K. In 2018, Sergei Skripal, a former Russian spy whom British intelligence managed to recruit as a double agent, and his daughter Yulia were poisoned with a Novichok agent that was applied to the doorknob of their home. A passerby found the Skripals incapacitated on a park bench and alerted police. When doctors in the hospital discovered Sergei's history, and considered both Skripals' symptoms, they began to suspect that a nerve agent had been used and initiated treatment.

Antidotes for poisoning by nerve agents have been extensively investigated. Since the 1930s, the first line of defense after exposure has been the injection of atropine, a drug derived from the belladonna plant. It is named after Atropos, the goddess in Greek mythology who decided when mortals would die by cutting their thread of life. Her sister, Clotho, spun the thread, and another, Lachesis, measured its length.

Atropine is known as an acetylcholine antagonist because it dislodges acetylcholine from receptor sites and thus reduces the risk of overstimulation. Atropine injection by itself is effective only for a short time since acetylcholinesterase remains inactivated by the nerve agent, so the concentration of acetylcholine will keep increasing and eventually overpower the protective effect of atropine. A second substance, called pralidoxime chloride, must be administered to release the nerve gas from the enzyme and destroy it. The patient may still be left with convulsions, which can be treated with diazepam (Valium). Exactly what else may have been involved in the treatment of the Skripals is not known, but both survived and were given new identities and are believed to be living in New Zealand. Alexei Navalny survived, returned to Russia and was arrested, causing an international furor. Dawn Sturgess was an unlucky, unintended victim of the Russian scheme.

Dawn Sturgess was not as lucky. Soon after the Salisbury incident, she was poisoned when she sprayed what she thought was perfume on her wrist from a bottle her partner had found in a garbage bin. This was no perfume. The bottle had been made to look like perfume but was actually a very sophisticated container that had been designed for two Russians to sneak a Novichok nerve agent into the U.K. The pair were later caught on closed-circuit TV and identified as secret agents. Sturgess was exposed to a much higher dose than the Skripals and the chemical was applied to a thinner, more permeable area of skin. Unfortunately, prompt medical treatment was unable to prevent her death. Strurgess's partner, Charlie Rowley, reportedly spilled some of the contents of the perfume bottle onto his hands, but immediately washed off the oily residue, which saved his life. Of course, the Russians, in spite of overwhelming evidence, deny involvement in any of these poisonings and even deny the existence of Novichok agents. World leaders such as Angela Merkel and Boris Johnson called for answers about the poisoning from Russian President Vladimir Putin, but U.S. President Donald Trump was silent on the subject.

NATURAL OR SYNTHETIC?

There are many confrontations on the battlefield of cyberspace. Vaccine proponents versus anti-vaxxers. Creationists versus evolutionary biologists. Anti-fluoride activists versus fluoridation supporters. Climate change deniers versus climate scientists. However, one of the most vigorous debates revolves around the relative merits of "synthetic" and "natural" chemicals. Worshippers of natural substances maintain that these are preferred over synthetics because they are safer since "nature knows best." Scientists on the other hand claim that the safety and efficacy cannot be determined by a substance's origin, but only by appropriate testing. They commonly point out that nature is replete with toxins ranging from strychnine and botulin to arsenic and snake venom.

The idea that natural substances have some sort of "vital force" that cannot be replicated in the lab was dismissed as early as 1828, when Friedrich Wöhler synthesized urea and showed it to be identical to the natural version isolated from urine. Nevertheless, the myth that there is something magical about natural substances persists to this day. Recently, I had an inquiry from an individual who was perturbed by learning that the caffeine in his energy drink originated in a lab and not in a coffee bean. I explained that a molecule of caffeine is defined by its molecular structure and whether the atoms that make up that structure are joined together by Mother Nature or by a chemist makes no difference. To the body, they are treated exactly the same because they are identical.

I thought I had provided a pretty convincing argument. But then came a follow-up question. "So, how come I saw this report about chemists finding that the caffeine in an energy drink didn't come from the coffee bean but was synthetic? If there is no difference, how did they know this?" Good question. While it is true that in terms of molecular structure natural and synthetic caffeine are identical, there is actually a subtle difference in isotope ratios that allows for the identification of a sample as being natural or synthetic.

An element is defined by the number of protons in its nucleus. For example, every atom of carbon in the universe has six protons in its nucleus. But the nucleus can also contain neutrons, particles that have the same weight as protons but have no effect on the identity of the element. Atoms of the same element that differ in the number of neutrons are called isotopes. Roughly 99 percent of all carbon atoms have six protons and six neutrons and are referred to as carbon-12. About 1 percent have seven neutrons and are therefore labeled as carbon-13. A tiny, tiny fraction, one in a trillion carbon atoms, has eight neutrons. This isotope, carbon-14, is radioactive, meaning that one of its neutrons breaks down into a proton and an electron, and by virtue of now having an extra proton, it becomes an atom of nitrogen. The emitted electrons constitute beta rays and can be detected. In any sample containing C-14, after some 5,730 years, half the atoms will have decayed. This is the basis of radiocarbon dating as well as identifying a sample as being natural or synthetic.

When neutrons in cosmic rays emitted by stars knock a proton out of nitrogen atoms in the atmosphere, they produce an atom of carbon-14. This carbon-14 then reacts with oxygen to form carbon dioxide, which is then taken up by plants during photosynthesis and used in the biosynthesis of all the plant's organic components. Therefore caffeine in a coffee bean will have some C-14. As long as a plant is alive, it keeps taking up carbon dioxide, so its C-14 content is the same as that of the atmosphere. Once the plant dies, it no longer photosynthesizes, and its C-14 content will continue to decrease through radioactive decay. By the time plant matter becomes petroleum, its C-14 content will have virtually disappeared.

Caffeine can be synthesized in the laboratory from simple molecules that are sourced from petroleum, and this version will have essentially no C-14. However, caffeine from coffee beans will have some, and its presence can be detected either by measuring beta ray emission or through mass spectrometry, an instrumental technique that can detect small changes in atomic mass.

Commercially, caffeine can be obtained either from the decaffeination of coffee beans or through chemical synthesis. In general, synthesis is cheaper, and most of the caffeine that is added to energy drinks is produced on a massive scale in China. Once produced, it is the same as any other caffeine. Well, almost. While the difference in C-14 content is only of academic interest, there may be a difference in residues of processing chemicals. In any chemical reaction the product will contain traces of the reagents used in its synthesis. There have been questions raised about quality control in some Chinese facilities. However, it is also possible that in the case of caffeine obtained by decaffeinating coffee beans, traces of extraction solvents remain.

Basically though, when it comes to the consumption of caffeinated beverages, the question that should be asked is not whether the caffeine is natural or synthetic, but rather what impact the world's most widely consumed psychoactive substance may have on health.

TIME TO SMELL THE ROSES

There is a classic victory celebration at the end of a Grand Prix Formula 1 race. The first three finishers mount the podium, then the winner pops a bottle of champagne, shakes it up, and proceeds to drench everyone around with the bubbly. Except when the race is held in Bahrain or the United Arab Emirates! In these countries, the public consumption of alcohol is not allowed, and in keeping with local culture and tradition, a fizzy beverage with a base of rose water is substituted for champagne. Rose water has a special place in Muslim culture, with the rose sometimes being referred to as the "Flower of Prophet Muhammad" as its pleasant aroma is said to be a reminder of the Prophet.

The simplest form of rose water is produced by steeping rose petals in water. This certainly smells like roses but does not have exactly the same bouquet as a freshly picked flower. Rose fragrance is a complex mixture of dozens of compounds, some of which are water-soluble, others not. For example, geraniol, a major component of rose aroma, is water-soluble, but beta-damascenone, another important contributor, is not. Nevertheless, the ancient Romans were very fond of such simple rose water, particularly at orgies, where the scent supposedly inflamed passions.

Much more fragrant rose water became available with the introduction of the alembic, a basic distillation apparatus, generally attributed to the Islamic alchemist Jabir ibn Hayyan in the ninth century. The alembic is a rounded flask with a long neck that to this day is identified with the practice of alchemy. When a substance placed in the flask is heated from below, the vapors formed travel through the neck whereupon cooling they condense to a liquid that drips out the end. If water is also added, the steam that forms helps to vaporize the volatile components. With rose petals and water in the alembic, the product of such steam distillation is the essential oil of the rose, which ends up floating on top of an aqueous layer, the hydrosol that contains the water-soluble components. It is this layer that is generally referred to as rose water.

The essential oil, also known as attar of roses, is used in perfumery, while rose water can be used to flavor beverages or sweets such as marzipan and Turkish delight. Rose attar can also be produced by agitating the petals in a vat with a solvent such as hexane. Such solvent extraction draws out the fragrant compounds along with waxes and pigments. The solvent can then be removed under vacuum and recycled. Treating the waxy residue with alcohol dissolves the fragrant components, and evaporating the alcohol under low pressure leaves behind the essential oil, or absolute. It can take more than two thousand flowers to produce a gram of oil, which means that rose attar is very expensive. This invites adulteration, generally by dilution with oil of geranium, which is rich in geraniol but much cheaper. There is no health issue here, but such "extended" oils should not be referred to as pure attar of roses.

It isn't surprising that a popular flower like the rose should have invited investigation for possible medicinal properties over the ages.

The ancient Greek physician Dioscorides recommended an elixir of roses cooked in wine for headaches, while Indigenous Americans treated colds, coughs, and fevers with various potions derived from the flowers, leaves, or roots of the plant. Rose water has also been used cosmetically, incorporated into creams for its supposed anti-inflammatory properties.

Such traditional uses, while mostly anecdotal, have stimulated some serious research into potential medicinal properties, especially since roses do contain numerous terpenes, glycosides, flavonoids, and anthocyanins with potential pharmacological effects. Unfortunately, virtually all the studies have been carried out in animals or in cell culture with a paucity of human data. For example, aromatherapists have claimed that rose essential oil can have a soothing, sleep-inducing effect. However, the citations provided deal with mice. When mice are put to sleep with a barbiturate, their sleeping time increases if their food is supplemented with some rose extracts. Only specific extracts work, and only at doses far greater than any to which humans would be exposed. Furthermore, the extracts were ingested, not inhaled.

Rose extracts have also been claimed to have a pain-killing effect. What is the evidence? When mice are treated with an ethanolic extract of roses and then are placed on a hotplate, they flick their tails less frequently than in the absence of the extract. Hardly enough to warrant a prescription for humans. As far as purported antimicrobial effects go, rose essential oil has been shown to be effective against a variety of microorganisms in a petri dish. There is nothing surprising here, since numerous plant extracts show such effects, but that is a long way from showing an effect in human infections.

While roses may have no practical medicinal effects, their visual splendor and striking fragrance have come to be associated with affection and an appreciation of the beauty of life. In these trying times filled with racial strife, political rancor, and daily death statistics, sniffing a little rose fragrance sounds like a pretty good idea.

LEGO CHEMISTRY

They foster imagination, help develop hand skills, and in these COVID days, keep kids from sitting in front of the TV all day. Lego sets began to entertain children with cleverly constructed interlocking plastic bricks way back in 1949, eventually expanding to highly sophisticated kits containing thousands of pieces that can be assembled into models of the Taj Mahal or the Apollo Saturn V launch vehicle. A far cry from the simple set of bricks, windows, and doors I grew up with.

Lego was founded in 1934 by Danish carpenter Ole Kirk Christiansen, the name being a contraction of the Danish phrase "leg godt" meaning "play well." At first Christiansen made wooden toys but switched to plastics in 1947 when advances in polymer technology made these materials readily available. In 1949, he heard about a toy that was being produced in Britain under the name Kiddicraft Self-Locking Bricks, invented by Hilary Fisher Page. These miniature bricks were made of injection-molded cellulose acetate and caught Christiansen's fancy. He purchased a molding machine and introduced what he called Automatic Binding Bricks. Page had patented his product, something of which Christiansen was apparently unaware. This led to some legal issues, eventually ending up in an out-of-court settlement with Lego acquiring the rights to the toy.

Christiansen was dedicated to producing a high-quality product and introduced the motto "only the best is good enough." Although cellulose acetate produced an acceptable product, it eventually lost its rigidity and limited the locking ability of the bricks. Intensive research resulted in a switch to acrylonitrile butadiene styrene (ABS), a polymer that is strong, isn't likely to warp, and is resistant to color-fading. ABS was born out of research by German chemical companies in the 1930s in a search for synthetic alternatives to rubber should the country be cut off from imported supplies of natural rubber in case of war.

Unlike cellulose acetate, ABS is opaque and cannot be used to make transparent pieces. These are now made from polycarbonate, a plastic formulated with bisphenol A (BPA) as a component. BPA has received a great deal of attention as a potential endocrine disruptor; however, once incorporated into the polymer, it is no longer present as such and does not pose a risk to children handling the relatively few transparent Lego blocks.

The tires on Lego vehicles are made from a different plastic since some flexibility is required. Styrene-butadiene-styrene (SBS) fits the bill. This polymer was also developed in Germany in the 1930s and is a "thermoplastic elastomer," meaning that it behaves like rubber at room temperature but can, like plastics, be molded into useful shapes such as tires, soles of shoes, conveyor belts, gaskets, and floor coverings when heated. Amazingly, Lego is the largest producer of tires in the world, albeit the tires are pretty small. The last reported production number by Lego is 318 million tires a year, which means some 36,000 being produced every hour. That is more than twice what Bridgestone, the leading automobile tire manufacturer, produces.

Lego churns out more than 60 billion parts every year, and that requires a lot of plastic. All the raw materials to make these plastics are sourced from petroleum, which is a nonrenewable resource. Recognizing this, the company has set a goal of making its products from sustainable sources by 2030, a significant challenge. The first step has been taken by introducing polyethylene made from ethanol derived from sugar cane production. While this plastic can be used to make the miniature trees and bushes that come in many Lego sets, it doesn't have suitable properties to make the interlocking bricks. And while sugar cane is indeed renewable, it does require the application of fertilizer and pesticides, and ethanol production requires quite an input of energy.

In the past, Lego has had some issues with the dyes used, especially yellow cadmium sulfide and red cadmium sulfoselenide, since cadmium compounds are toxic. These dyes were used until about

1980, when a switch was made to less toxic perinone and azo dyes. However, the amount of cadmium that leached out of Lego bricks was never significant enough to cause a hazard to children.

Unfortunately, some Lego products do end up being improperly discarded and end up as environmental pollutants. In a recent study, Lego bricks that washed up on beaches in England were examined and found to have lost mass, meaning that they may be a source of microplastics. When researchers examined the extent of the weathering of these bricks, they concluded that Lego plastics can last in the ocean for hundreds of years.

It is a shame that some of these toys end up in the ocean because they should never be discarded. There is always a child somewhere who would be happy to spend hours building some imaginative structure. It is increasingly important to prevent Lego parts from ending up in the environment because every second, more and more pieces are being produced. One Lego creation in 2020 was a 2,646-piece reproduction of the original Nintendo console and retro TV set featuring Super Mario Brothers on the screen. What a way to recapture the childhood magic of those COVID-free days! Just make sure you never step on a piece barefoot. That can also be a memorable experience.

CHEERS TO THE DISCOVERER OF WINE DISEASE

The epidemic in 1639 rocked the region of Poitou in France to its core. As described by Francis Citois, Cardinal Richelieu's personal physician at the time, many of the "Pictones," as the area's inhabitants are known, experienced symptoms starting with colic (abdominal pain), fever, lethargy, constipation, and insomnia before progressing to visual problems, mental deterioration, paralysis, and often death. The cause of "colica Pictonum," as the disease came to be called, was unknown. There was nothing unusual about that since medical knowledge was primitive and illnesses were generally attributed to an imbalance in the body's "humors," divine retribution for sin, or spells cast by witches.

Citois was not the first to describe such cases. Reports of severe colic, followed by loss of control of the extremities, blindness, and insanity, can be traced back to ancient Rome. Then in the eleventh century, Avicenna described an epidemic of these symptoms, as did Paracelsus in the sixteenth, and English physician Thomas Sydenham in the seventeenth century. But it remained for an obscure German physician, Eberhard Gockel, to finally discover the root cause of colica Pictonum. Gockel was not an ideal candidate for making a major medical discovery given that he was the author of a book about illnesses caused by witches and werewolves, and a treatise on a rooster that reputedly laid an egg. In spite of such follies, Gockel, as "City Physician of the Town of Ulm," managed to come to a correct conclusion based on observations he made in a couple of monasteries that were under his care.

Gockel regularly visited the monasteries during an outbreak of colica Pictonum. On one such visit, he was offered a glass of wine, whereupon he was "attacked by the most atrocious colic pains and terrible fever." Since these symptoms mimicked the disease he witnessed in a number of the monks and guests at the monastery, his attention turned to the wine. A visit to the wine cellar yielded a clue. The bottom of the barrel, which had been purchased from a wine merchant in nearby Goppingen, contained a viscous sediment!

Gockel tracked down the merchant and learned that he routinely added a concoction of powdered litharge steeped in vinegar to the wine as a preservative and sweetener, especially when unsuitable weather conditions resulted in a grape harvest that led to sour wine. Litharge is a natural mineral form of lead oxide, and its addition to wine had been a common practice since Roman times but had not previously been connected to disease. Now it aroused Gockel's suspicion. His hunch was placed on a firmer footing when he learned that two visiting friars had contracted colic after sharing their hosts' wine and had recovered completely upon returning to their own monasteries. When questioning revealed that none of the monks who abstained from wine had contracted the disease, Gockel was sure he had found the cause. Colica Pictonum was poisoning caused by lead.

In 1697, Gockel published his account of "wine disease" in a paper entitled, "About the sweetening of acid wine with litharge with the greatest harm to those who drink it." In it, he gave credit to Samuel Stockhausen, who may well deserve the title of being the father of occupational medicine. In a book published in 1647, Stockhausen discussed the short life expectancy of miners, a phenomenon that had been known since ancient times. He paid particular attention to those involved in the mining of lead, and the symptoms he described were recognized by Gockel as being identical to what he had observed in his adulterated wine drinkers. This helped cement Gockel's theory about colica Pictonum, and its various historical manifestations were now seen as the result of consuming wine that had been doctored with lead.

The Romans had accidentally discovered that wine stored in lead vessels was less prone to spoilage and, as a bonus, developed a sweeter taste. This sparked the idea of boiling grape juice in a lead pot to reduce its volume and concentrate the sweetening effect. The resulting syrup was known as "sapa" or "defrutum" and was added to wine, particularly if the wine was sourer than desired. This practice was continued throughout history, often referencing a treatise by the Roman agricultural writer Lucius Columella, who gave specific instructions for the preparation of sapa. The lead content of wine prepared according to Columella's directions has been estimated to be about 20 milligrams per liter. Since 0.5 milligrams of lead per day can produce chronic lead poisoning, it is no surprise that until Gockel's discovery, oenophiles were plagued with colica Pictonum.

The Ulm tainted-wine episode led to the passage of a law against the production of leaded wines in Germany. At first, not much attention was paid to the law, but when a cooper, Johann Ehrni, was beheaded in Stuttgart after being found guilty of "correcting" his wines with litharge, the practice came to a dead stop.

Eberhard Gockel is not one of the famous names in the history of medicine, but he deserves a great deal of credit for shedding light on the misery of colica Pictonum through his observation of teetotaling monks. Gockel himself did not exactly lead a monastic life — he fathered eighteen children! As a final footnote, the wine dealer who had sold the leaded wine to the monastery broke his neck when he was thrown by his horse. Some saw this as divine retribution.

DIRTY BOMBS AND PRUSSIAN BLUE

I think *Goldfinger* is the best of the James Bond movies. Remember the scene when Bond is secured to a table made of pure gold and Goldfinger activates a laser that starts to melt the metal and threatens to bisect 007? "Do you expect me to talk?" Bond asks. "No, Mr. Bond, I expect you to die!" Of course, Bond manages to escape and goes on to foil Goldfinger's scheme of exploding a "dirty bomb" in Fort Knox to render the U.S.'s gold supply radioactive and unusable, thereby increasing the value of his own stock of gold.

That was before global terrorism introduced the possibility of inflicting radioactive damage on a population with a device that combines radioactive material with conventional explosives, a so-called dirty bomb. This, of course, requires the acquisition of radioactive material — not easy, but possible. There are a number of radioactive isotopes that are produced in nuclear reactors that have practical application. Cobalt-60 is used in the radiotherapy of cancer, americium-241 is used in smoke detectors, and iridium-192 is used in industrial radiography to locate flaws in metal components. Terrorists could conceivably get their hands on such radioactive isotopes.

Much research has been dedicated to finding antidotes for radioactive materials that may be inhaled or ingested in case of a nuclear accident or terrorist attack. The most publicized antidote has been the swallowing of potassium iodide pills with the intent of saturating the thyroid gland with iodide and preventing the uptake of radioactive iodide that may be released in a Chernobyl-type accident. Another possibility that has emerged is a pigment known as Prussian blue, available under the name Radiogardase. Encased in gelatin capsules, the blue powder would be swallowed in case of radioactive substance exposure. Prussian blue has the ability to bind the radioactive metal ions that are released by dirty bombs and subsequently eliminate them from the body. It can also be used in cases of thallium poisoning.

This pigment has a fascinating history. The name may ring a bell for people who remember Prussian blue as one of the original Crayola colors. The name was changed to midnight blue in 1958, supposedly because the company thought people were mystified by the term "Prussian." In any case, Prussian blue was actually the first synthetic pigment ever discovered, predating mauveine, and the first synthetic dye. Dyes are colored soluble chemicals that are absorbed into the material to which they are applied, while pigments consist of extremely finely ground insoluble particles that coat a surface over which they are spread. Like William Henry Perkin's famous accidental discovery of mauveine, the discovery of Prussian blue was also serendipitous.

In 1706, Johann Jacob Diesbach, a color merchant, needed help in producing a red dye from cochineal insects and consulted the philosopher and theologian Johann Konrad Dippel, who had a history of dabbling in alchemy. Dippel sought not only to turn base metals into gold, he also searched for the "elixir of life," a potion that would enhance longevity. This, he believed, could be extracted from animal parts, particularly bones, after these had been decomposed by the addition of potash. Distillation of the mixture yielded a foul-tasting, malodorous oil that came to be known as Dippel's Oil. Anything that tasted and smelled so bad had to be good for you! Dippel's fame spread, and his universal cure flourished for a hundred years.

Herr Diesbach had some experience with making a red dye from the cochineal insect by cooking the little bugs up with green vitriol and potash. But now he was having trouble finding potash. The substance referred to as green vitriol was iron sulfate, while potash was the "ash" left behind when a mixture of wood residue and vegetable matter was boiled to dryness in a pot. It was mostly potassium carbonate. Dippel, of course, had plenty of potash. After all, it was a key ingredient in his magical remedy. But when Dippel's potash was mixed with Diesbach's vitriol, the results were absolutely startling. A beautiful blue color was formed! It turned out to be just the right hue for the uniforms of the Prussian army and came to be known as Prussian blue. Artists also took to it. The sky in Vincent van Gogh's famous *The Starry Night* owes its stunning blue color to Prussian blue. The pigment quickly replaced the naturally occurring aquamarine, which was very expensive given that it was derived from emeralds!

Neither Dippel nor Diesbach understood the chemistry of their accidental discovery. It seems the potash was contaminated with naturally occurring ferrocyanide that combined with iron sulfate in Diesbach's cochineal solution to yield the novel blue. This chemistry was supposedly used by German spies during the Second World War to write secret messages. A solution of ferric sulfate can be used as invisible ink that turns blue when sprayed with ferrocyanide.

And there is one final twist to this story. Dippel supposedly also carried out dissections and experimented with transferring "souls" between cadavers with a funnel. In 1816, Mary Shelley, traveling with her husband Percy Bysshe Shelley along the Rhine, supposedly visited the castle where Dippel had carried out his anatomical experiments. The name of the castle? Castle Frankenstein. And the rest, as they say, is history.

A SHOCKING CATASTROPHE IN BEIRUT

The Beirut explosion may be the worst ammonium nitrate disaster in history, but it is not the first such calamity. Let's go back to the early morning of April 16, 1947, when spectators flooded to the docks in Texas City, Texas, drawn by the bright orange flames and the massive plume of black smoke that enveloped the SS Grandcamp, a French ship that had caught fire in the harbor. Then, as people marveled at the inferno, and quick-thinking vendors circulated with peanuts and other refreshments, there was a reverberating explosion. Hot pieces of metal from the disintegrated ship rained down, a devastating shock wave rolled across the land and sea, and within minutes much of Texas City was in flames. Almost six hundred people perished, many of them the onlookers who had come to gawk at the spectacle.

What cargo was responsible for the disaster? Nitroglycerine? TNT? Dynamite? None of the above. It was fertilizer! Not any old fertilizer, mind you. The Grandcamp had been loaded with 2 million kilos of ammonium nitrate destined for Europe. Ammonium nitrate is rich in nitrogen and can yield bumper crops or green up a lawn. But it can also explode and cause terrible bloodshed. An explosion can best be described as a "sudden going away of things from the place where they have been." The cause of such swift departures is a shock wave formed by the very rapidly expanding gases that characterize an explosion. In the case of ammonium nitrate, the gases are water vapor, oxygen, nitrogen, and oxides of nitrogen. Don't get the impression though that ammonium nitrate explodes easily. It doesn't. Various conditions have to be met for an ammonium nitrate explosion to occur.

Let's get back to the SS Grandcamp. A fire broke out in the hold, most likely due to an improperly discarded cigarette. Fearing damage to his cargo, the captain decided not to try to extinguish the flames with water. Instead, he ordered the hatches to be battened down, hoping to cut off the fire's oxygen supply. It didn't work and the cargo of ammonium nitrate began to heat up. At first, it just decomposed into steam and nitrous oxide, better known as laughing gas. But it was no laughing matter when the high temperatures triggered the breakdown of the laughing gas into nitrogen and oxygen. The fire, now well supplied with oxygen, intensified. Still, there probably would have been no explosion had it not been for two other factors. The ammonium nitrate was packed in paper bags which began to burn with great intensity, and more significantly, the ship had been filled with 1,500 tons of fuel oil. When the oil caught fire and its hot vapors mixed with the ammonium nitrate, which by now was venting massive amounts of oxygen, the conditions were right for "a very sudden going away of things from the place where they had been!"

The Texas City disaster was an accident. But the bombing of the Federal Building in Oklahoma City in 1995 certainly was not. The chemistry, however, was the same. The perpetrators of this horrific crime were aware of the explosive nature of mixtures of ammonium nitrate and fuel oil (ANFO). Indeed, most commercial explosives in North America used in mining and construction fall into this category. ANFO mixtures are actually remarkably safe to use; they have to be detonated by an explosive charge. Of course, Timothy McVeigh and his cronies had no access to commercial explosives, so they decided to make their own. Either they researched their subject remarkably well, or they were very lucky. Homemade brews of fertilizer-grade ammonium nitrate and fuel oil are very difficult to detonate without the use of TNT, dynamite, or blasting caps. Unfortunately, as history has shown, terrorists can be remarkably resourceful.

The Beirut explosion in all probability was an accident, but exactly how the ammonium nitrate ignited may never be determined. Certainly, there was negligence involved. Such massive amounts of the chemical should never have been stored in such a fashion. However, as far as the ammonium nitrate fertilizer you may have purchased from a garden supply store goes, the chance that it will spontaneously blow up is roughly zero.

THE STORY OF VODKA

As the story goes, the world's best-selling vodka, Smirnoff, owes its original fame to a clever marketing gimmick by Pyotr Smirnoff, a Russian serf who by the time of his death in 1898 had become one of the wealthiest men in Russia. And it was all thanks to vodka.

Smirnoff did not invent vodka, a beverage that is composed of water and alcohol and essentially nothing else. The origins of this spirit date back to the invention of the still by Arab alchemists around the eighth century. Distillation allowed for the isolation of pure alcohol from a solution produced by the fermentation of some starchy or sugary grain, fruit, or vegetable. Although people commonly associate vodka with potatoes, corn or wheat are much more commonly used.

Fermentation is a process whereby a microbe, such as a yeast or bacterium, releases enzymes that convert starch or sugar into alcohol. Once pure alcohol was obtained, it could be combined with water in any proportion. Originally, such alcoholic blends were used as medicine, but by the fourteenth century, vodka, the term originating from the Slavic *zhiznennia voda*, meaning "water of life," was being consumed as a beverage in Russia and Poland. "Voda" was eventually changed to the affectionate nickname "vodka," meaning "little water." Since the drink contained "little water" and lots of alcohol, it readily led to a state of alcoholic stupor, a problem that has plagued Russia to this day.

By the end of the eighteenth century, there were a number of distilleries in Russia that produced increasingly pure vodka, especially after St. Petersburg chemist Theodore Lowitz demonstrated that filtration through charcoal removed residual impurities. Market competition was fierce, and Pyotr Smirnoff, who had founded a small vodka operation, had an idea to get a foot up on his rivals. Actually, a bear's foot.

At a fair, Smirnoff exhibited a couple of bears he had trained to drink vodka from a glass they held with their front paws while waiters dressed as bears served Smirnoff's vodka to the gathering. The real bears actually did knock back the vodka, rendering them sleepy and amiable throughout the fair. Supposedly, the czar visited the fair and was taken by the display. He gave the vodka his benevolent approval and sales took off!

By 1904, Vladimir, Pyotr's son, had taken over the company and was producing more than 4 million cases of vodka per year. The

Smirnoff family became fabulously rich and the czar, on seeing the money that could be made by the sales of vodka, decided to nationalize the industry, forcing Vladimir to sell his factory and brand. In 1917, the Smirnoffs, now considered aristocrats, had to flee Russia and ended up settling in France where Vladimir founded a distillery once more. Actually, this is when the original name "Smirnov" was changed to Smirnoff, the contemporary French spelling of the name. In 1933, Vladimir sold the rights to produce his vodka to Rudolph Kunett, a Russian who had immigrated to the U.S., and with that the American vodka industry was born.

The popularity of the beverage is due to its high alcohol content and general lack of taste, which makes it ideal as the base for many a cocktail. The Bloody Mary, Moscow Mule, and of course the vodka martini (shaken, not stirred, as James Bond prefers) are examples. You can even make "Vodka Gummy Bears" by soaking the gummies in vodka for three days in a glass bowl in the fridge. Apparently, sugar-free versions maintain the texture better. I bet these bears can add quite a punch to a party, just like Smirnoff's bears once did.

Vodka is 80 proof, meaning it is 40 percent alcohol by volume. That is not high enough to inactivate viruses, so vodka should be poured into a glass, not onto your hands. Contrary to a widely disseminated myth, this 40 percent standard was not set by the great Russian chemist Dmitri Mendeleev, who is best known for his formulation of the periodic table of the elements in 1869. The myth can be traced to an advertisement by Russian Standard, a popular vodka that claims: "In 1894, Dmitri Mendeleev, the greatest scientist in all Russia, received the decree to set the Imperial quality standard for Russian vodka and the 'Russian Standard' was born." This is not true.

Mendeleev's 1865 doctoral thesis was entitled "A Discourse on the Compounds of Alcohol and Water" and dealt with the density and thermal expansion of a mix of alcohol and water at various ratios. It had nothing to do with vodka. Mendeleev did sit on a Russian commission that examined ways of taxing alcohol efficiently, but again this was not specific to vodka. He had nothing to do with the 40 percent alcohol content of vodka, but nevertheless the story has even given birth to an American vodka called "Mendeleev." The father of the periodic table would not have approved. The great scientist did not drink vodka for fear of becoming an alcoholic like one of his brothers. He preferred wine.

GRADUATING TO BIOPLASTICS

In the classic film *The Graduate*, a family friend corrals young Benjamin, who had just graduated from college, and whispers into his ear: "Plastics." The year was 1967 and plastics were the "miracle materials" that changed lives. Virtually every industry from airplane and car manufacturers to hospital equipment and cookware suppliers benefited from plastics. Women flocked to Tupperware parties and vinyl records were the rage. My, how times have changed! Plastics have become a pariah, pollutants that are sullying and endangering the environment. If Benjamin were embarking on a career today, the whispered word would likely be "bioplastics," with further qualifiers such as "degradable," "biodegradable," "oxo-degradable," "recyclable," or "compostable."

These terms have infiltrated advertising as the plastics industry attempts to cope with the negative image of beaches defiled with plastic bottles, garbage patches in the ocean, shopping bags in gutters, and straws up turtles' noses. Indeed, items made from the most common plastics, such as polyethylene (PE), polypropylene (PP), polyethylene terephthalate (PETE), and polyvinyl chloride (PVC), are extremely durable and can stay around in the environment for decades and decades.

That polyethylene hula hoop purchased back in the 1950s looks none the worse for wear, and that plastic bottle carelessly tossed into the ocean from a boat may float for decades before being battered enough by waves to break into smaller pieces that can then be mistaken for food by fish and end up in their stomach, and eventually, in ours.

Most plastics are made from natural gas or petroleum, which is another concern since these are nonrenewable resources. So, we have a dual challenge. Find ways to cut down on the amount of plastic that gets discarded and find renewable resources that can be converted into bioplastics.

Bioplastics are produced partly or wholly from living species such as plants or microbes rather than from fossil fuels. An example would be polylactic acid (PLA), made from lactic acid, which in turn is produced by the action of lactic acid bacteria on starch from corn, cassava, sugarcane, or sugar beets. PLA is deemed to be a "green" plastic because it is made from a renewable resource. It can be used to make transparent drinking cups, disposable tableware, and teabags.

However, bioplastics are not necessarily biodegradable. That term refers to substances that can be broken down by microbes normally found in the environment. Food scraps, human and animal waste, paper, wool, and fallen leaves are biodegradable, but the speed at which this happens depends on the conditions. A cup made of PLA will biodegrade quite quickly if buried in soil but will not break down in a landfill. It is, however compostable, which is why it is used to make compost bags. Note, though, that compostable does not mean that the plastic will break down in the compost heap in your backyard. "Compostable" on a label means that in a proper composting facility where temperature and pressure are strictly controlled, it will decompose into carbon dioxide, water, and a complex mixture of organic compounds such as cellulose, hemicellulose, and lignin, referred to as biomass. This will not happen in a landfill.

Furthermore, some bioplastics may not be biodegradable or compostable at all. For example, polyethylene can be made from ethylene that is made from ethanol that in turn is produced by fermentation of corn or sugar cane. This polyethylene is identical to the polyethylene made from ethylene derived from natural gas or petroleum and is not biodegradable or compostable, but could still be advertised as a "bioplastic" or as "renewable polyethylene." Does it have any advantage over regular polyethylene? Maybe. When the corn or sugar cane is growing, it uses up carbon dioxide from the air through photosynthesis. On the other hand, the land dedicated to the crops requires deforestation and the use of fertilizer and pesticides.

While polyethylene is not biodegradable, it is recyclable. But you have to remember that recyclable does not necessarily mean recycled. Unfortunately, much of the plastic that goes into the blue bins ends up in landfills or incinerators. Separation of plastics is a big problem for the recycling industry. Biodegradable and compostable plastics are difficult to separate from other plastics and can contaminate the recycling stream and should not be put in the recycling bin.

The term "degradable" on a label is misleading. It just means that in some unspecified time, when exposed to the elements, it will break down into smaller pieces. A plastic bag can be said to be degradable because if it happens to end up in the ocean, it will eventually break down into smaller pieces, but those smaller pieces of plastic can be a problem. "Oxo-degradable" means that metallic catalysts have been added to the plastic to cause a more rapid degradation when exposed to heat or light, but the breakdown products may not be biodegradable, so this is not of great advantage.

There is no simple solution to the plastic waste problem. But we can cut down on use, although that is not easy. I have been reusing a shopping bag for a year, but that also means I sometimes have to buy garbage bags. And remember that it is possible to drink a soft drink without a straw. So, forget the straw. Better yet, forget the soft drink.

SPIRITS OF SALT

Constantine Rhodocanaces had come to England sometime in the seventeenth century from the Greek isle of Chios and was brewing up medicine "in a Physicall Laboratory in London, next door to the Three Kings-Inne." I know this because it says so right on the cover of a booklet he published in 1663. A booklet that I now have! Not a copy, not a facsimile, but the original!

I have a penchant for unusual historical items, especially if they deal with chemistry or pseudoscience. My interest is aroused even more if these two coalesce. That is just the case for Rhodocanaces's publication about his wonder product, the "Spirit of Salt of the World." I came across a reference to this epic when I was doing some research for my own recent book, *A Grain of Salt,* which also deals with chemistry and pseudoscience. When I saw that Spirit of Salt of the World was available from an antique book dealer, I jumped, and I now am the proud owner of this wonderful little relic.

The first line that strikes your eye on the cover is "ALEXICACUS," which is Greek for "averter of evil." And that is just what Spirit of Salt of the World promises to do. Avert all sorts of evil diseases from scurvy and "inflammation of the feet" to the "French Disease" and kidney stones. Rhodocanaces informs the reader that he works in a lab "where all manner of Chymicall preparations are carried on without any Sophistication or abuses whatsoever." This is where his Spirits of Salt of the World is "now Philosophically prepared and purified from all hurtful or Corroding Qualities, far beyond anything yet known to the World, being both safe and pleasant for the use of all Men, Women and Children." In those days, "sophistication" meant "deception" and "philosophically" was the term for our "scientifically." While there certainly was science involved in preparing the product, there was also deception.

So, what was Spirit of Salt of the World? Spirit of salt is an old term for hydrochloric acid. And that is what Rhodocanaces was peddling and claiming that it was a virtual panacea. Hydrochloric acid was well known at the time, having been discovered some eight hundred years earlier by the alchemist Jabir ibn Hayyan, who produced it by mixing salt with sulfuric acid. Ibn Hayyan is also credited for having discovered that distillation of a solution of a mineral we now know to be iron(II) sulfate heptahydrate yields sulfuric acid. Rhodocanaces was not the first to claim therapeutic properties for hydrochloric acid, but he claims his product is superior to others being sold as Spirit of Salt, which is outright dangerous. He tells us how he heard from the Right Worshipfull Thomas Middleton, apparently a trustworthy nobleman, that "not long since a sick person making use of the common Spirit of Salt bought at the Apothecaries, died upon the taking of it." Having heard this, "charity towards my Neighbor commandeth me to make publick that hereafter greater caution may be had in using the vulgar Corrosive Spirit of Salt, instead whereof I make publick this, which is most innocent, and healthful, as may be seen in the following testimonies."

And there are testimonies galore. Mrs. Bird gave some of Rhodocanaces's acid to her children who were troubled with worms, "which it presently kill'd and brought away." A man "sick of an inveterate Head-ach, which afflicted him at certain times every day, having been left by Physicians, and in opinion near death, did after purgation once and again prescribed, make use of this Spirit of Salt for the space of a week and was thereby suddenly and strangely recovered." Strangely indeed.

There is more. Rhodocanaces's concoction "procures a good appetite and prevents putrifaction of anything in the Stomak, prevents Drunkennesse and sickness therefrom, expels diseases that arise from corrupted blood, by purifying it, that lay before idle, and settled in the veins, and makes it volatile, and to proceed more regularly in its circulation." It also "keeps arteries from all filth, or slime, and sends away the water that lurks betwixt the skin and flesh, by Stool and Urine."

Of course, only the original will work. "There are some who pretend to make this Spirit according to my preparation, wherefore I think good to let the world know, that as yet this Secret hath not been communicated to any."

Why do I find this booklet so fascinating? Because it could have been written today. All you have to do is substitute "Spirit of Salt" with the name of one of the current supposed panaceas that entice the desperate and the worried-well with the same type of testimonials, smearing of competing products, and claims of secret breakthroughs. Of course, all those claims should be taken with "A Grain of Salt." Plus ça change, plus c'est la même chose.

UNEARTHING A GIANT HOAX

Goliath wasn't just a large man. He was a giant! According to the Bible, he was "six cubits and a span" in height, which translates to about ten and a half feet. And he is not the only giant to appear in the scriptures. In Genesis 6:4, for example, we learn that "there were giants in the earth in those days." Clearly then, any physical evidence that giants once really roamed the earth would be welcomed by biblical literalists. In October of 1869, it seemed that such evidence was unearthed. Literally.

William "Stub" Newell owned a farm in Cardiff, New York, just south of Syracuse. He needed a well dug and hired two men for the job, pointing out exactly where they were to dig. It wasn't long before their shovels struck what at first seemed like a large boulder. But further digging unearthed a giant stone figure of a naked man! Word quickly spread about the spectacular find. What was it? An ancient statue? Or perhaps the fossilized remains of a true giant? Newell was quick to capitalize on the discovery, charging people fifty cents to view the colossus. Thousands flocked to the farm, with visitors drawn into debates about the origin of the mysterious find. There were the petrificationists, mostly religious individuals who believed they were looking at the body of a petrified man and claimed that this was evidence of giants, just as described in the Bible. Others speculated that it was a stone carving dating back to some previous civilization. Newell didn't care what people thought as long as the money kept flooding in, which it did. The Cardiff Giant was attracting the curious and the faithful from across the country.

But the giant wasn't a petrified man, nor was it an ancient statue. It was only about a year old and was poised to become the central figure in one of the most successful hoaxes in history! George Hull was a tobacco farmer who had fallen on hard times and thought about going west to prospect for gold. He got as far as Iowa, where he encountered something that would turn out to be more precious than gold, human gullibility. The giant caper was stimulated by a chance meeting with a clergyman who espoused a belief in biblical giants. Hull was an atheist who thought this was outlandish and hatched a scheme that would demonstrate human credulity and simultaneously solve his financial problems.

Fossil finds were big news at the time, with the first dinosaur bones having been recently discovered. Hull was also familiar with petrified trees, so he thought that the discovery of a petrified man with Goliath-like proportions would create a lot of excitement among biblical literalists and would serve to poke fun at them when the hoax was revealed. He purchased a huge block of gypsum from a quarry in Iowa and had it hauled to a stonecutter in Chicago who was sworn to secrecy. The sculptor was asked to carve out the image of a twelve-foot-tall man with a somewhat contorted body. Using needles, Hull then riddled the statue with tiny holes to simulate pores and rubbed the gypsum with sand, ink, and acid to give it an antiquated appearance. Ink penetrated the gypsum where it had been dissolved by the acid and formed lines that gave the illusion of veins.

Hull then approached Newell, his cousin, who quickly agreed to burying the giant on his farm and scheming to have it revealed by well-diggers in return for a share of the profits. There was some concern that someone may have caught a glimpse of the burial, but when nothing came to light after a year, the diggers were hired and the hoax was on!

Scientists were also attracted to the find and quickly concluded that petrification, the process by which organic material becomes fossilized as the original tissues are replaced with minerals, was out of the question in this case. It didn't matter; people preferred to believe the unbelievable!

Hull eventually sold the giant to a group of businessmen headed by David Hannum, who then received an offer from P.T. Barnum, the celebrated showman and promoter of numerous hoaxes. When the offer to buy the giant was refused, Barnum had a copy made from plaster and papier-mâché and exhibited it in New York, claiming that it was the original! The king of humbug had made a fake of a fake and profited richly from it! When Hannum heard about this sham, he uttered the famous phrase "there is a sucker born every minute," which has since erroneously been attributed to Barnum.

By 1870, the giant ruse had been widely ridiculed by scientists and finally came to an end when the Chicago sculptor who had chiseled the "petrified man" confessed. Today, the Cardiff Giant can be seen at the Farmers' Museum in Cooperstown, N.Y., where it serves as a reminder of human gullibility, a phenomenon that is as prevalent today as it was at the time when the public was taken in by George Hull's giant hoax.

BISPHENOL A AND MY SOCKS

My sock drawer is over-brimming. I like interesting hosiery, especially when it features ducks. Never did I think that when I donned my socks I was exposing myself to bisphenol A (BPA), a chemical that, if you go by some bloggers who fancy themselves as experts in toxicology, was concocted by the devil himself.

I was alerted to the sock situation by a paper published in the journal *Environment International* that described snipping pieces out of newly purchased children's socks and testing them for the presence of bisphenol A as well as for estrogenic activity, both of which were found to be present. Furthermore, it turns out that other researchers have found BPA in pantyhose and various textiles. But why would scientists be interested in the chemical anatomy of socks in the first place?

Bisphenol A is one of the chemicals produced in the highest volumes in the world, to the tune of some 10 million metric tons a year. By comparison, this is 250 times more than the amount of aspirin that is produced. About 95 percent of all BPA goes towards the production of polycarbonate plastic and epoxy resins. Polycarbonate, thanks to its lightness, transparency, and toughness, is widely used in automobiles, airplanes, computers, cell phones, sports equipment, shatterproof labware, protective eyewear, casings for streetlights, bullet-resistant "glass," greenhouses, road signs, neonatal incubators, laparoscopic equipment, and dialysis machines. Epoxy resins find application as liners in food cans that prevent contact between the contents and the metal, preventing both corrosion and contamination. Epoxies also are found in quick-drying paints, boat hulls, and protective coatings on pipes and concrete floors. Bisphenol A is also used in dental sealants, as an antioxidant in some plastics, and in the production of thermal cash register paper.

Obviously, there is no doubt about this compound's practicality. But doubts have been raised about its safety. Bisphenol A has a molecular structure that partially resembles that of some steroid hormones such as estrogen and can therefore act as an "endocrine disruptor," either blocking or enhancing the effect of such hormones. Indeed, animal-feeding studies as well as experiments with cells in the lab have documented that BPA can cause reproductive and developmental problems, altered immune function, and hormone-related cancers. There have also been some human observational studies associating elevated blood or urine levels of BPA with obesity, diabetes, cardiovascular disease, prostate cancer, and impaired kidney, liver, and thyroid function. The animal studies have been criticized for using unrealistic doses and the fact that rodents do not metabolize BPA the same way as people. As far as human studies go, they can only show an association and cannot prove a cause-and-effect relationship. After taking the literally thousands of studies on BPA into account, most regulatory agencies have concluded that human exposure is indeed widespread but that the amounts finding their way into the body are unlikely to cause harm, especially given that BPA and its metabolites are relatively quickly excreted in the urine.

Nevertheless, there is the usual call for more research and substitutes that come with less baggage than bisphenol A. This is especially desirable for the production of epoxy resins for can linings, since it is the leaching of BPA from these that is responsible for the majority of human exposure. There are other materials such as acrylics and polyesters that can be used for some foods, but they do not have the universal applicability of epoxy resins made with BPA. That's why tinkering with the molecular structure of BPA is an attractive proposition, in hopes of coming up with a novel version that retains the benefits but eliminates the problems.

The commercial synthesis of bisphenol A involves reacting two molecules of phenol (hence "bisphenol") with one of acetone. The "A" is for acetone. The first substitutes that functionally acted like BPA were bisphenol S, made by combining phenol with sulfuric acid (the *S*) and bisphenol F that used formaldehyde (*F*) instead of acetone. These allowed producers to claim "no BPA" on labels, aiming to allay consumer fears. However, as it turns out, these compounds also have endocrine-disrupting properties, which isn't surprising since their molecular structure is very similar to BPA.

The question then comes down to making a compound that is different enough in structure from BPA so as to not be an endocrine disruptor, but similar enough so that it can engage in polymerization reactions like BPA. Chemists at Sherwin-Williams, a company that manufactures coatings, used a computer modeling program that allowed them to see how various derivatives of BPA would fit into hormone receptors. After scrutinizing many compounds in this fashion, they found one, tetramethyl bisphenol F (TMBPF), that seemed to be biologically inert but reactive enough to make a polymer.

Of course, computer modeling is not the same as a lab experiment. That's why the researchers sought the help of Tufts University's Dr. Ana Soto, a recognized expert in endocrine disruption, who put the chemical through a battery of tests and concluded that not only did TMBPF show no hormone-like activity in estrogen assays or rat studies, it did not migrate into food from a can coating at all. Problem solved? Maybe. To ensure no reproductive or developmental problems, multiple generations of animals will have to be studied and questions of cost and extent of applicability will have to be answered. What we have here, though, is a demonstration of how problems introduced by chemistry can also be solved by chemistry.

As far as my socks go, I'm really not worried about the minute traces of bisphenol A, especially since I do not plan to be dining on the socks. I suspect though that some activists may be keen on having me stuff one in my mouth.

TURNING UP THE HEAT ON THERMAL PAPER RECEIPTS

When you are spending money at a store, the cost may be more than the amount shown on the cash register receipt. According to some researchers, there is a cost to health. That's because handling the receipt transfers a chemical known to have hormone-disruptive properties to the skin, from where it can migrate into the bloodstream. That chemical is bisphenol A, commonly abbreviated as BPA. This is a multifunctional substance that is a component of polycarbonate plastic as well as the epoxy resin that lines food cans. In the case of receipts, it is coated onto the paper to the extent of about 20 milligrams per gram and acts to develop the image when heat is applied. There is some fascinating chemistry involved here. Leuco dyes are chemicals that can exist in a colorless or colored form depending on temperature and acidity. In this case, the paper is treated with the colorless form. When heat is applied, as directed by a printer head, the colorless form combines with BPA, here acting as an acid, to form the colored image.

A number of studies have shown that BPA can be transferred to the skin from thermal paper. This is done by having subjects handle the paper and then extracting their skin with a solvent such as ethanol and then testing the ethanol for bisphenol A content. Such studies have clearly shown that some of the chemical is transferred and that the transfer is significantly enhanced if previously a sanitizer or moisturizing cream had been applied to the hands. Following the handling of receipts, the concentration of bisphenol A in the blood and urine can also be measured. Indeed, some researchers believe that for most people, cash register receipts represent the most significant exposure to BPA.

The amount of BPA that shows up in the blood after handling receipts has been found to be more than if a comparable amount were consumed. That's because orally ingested BPA travels through the liver where it is metabolized, with the remnants being excreted in the urine. By contrast, transdermal passage does not lead to quick detoxication by the liver. There is also the issue that when BPA is transferred to the fingers, it can further contaminate other substances that are handled, such as food. In one study, eating French fries after handling cash receipt paper resulted in higher blood levels of BPA than after eating the fries with hands that had not touched such paper.

Of course, one cannot equate the mere finding of a chemical in the blood or in the urine with the presence of risk. Indeed, high urinary levels may mean that the chemical is being efficiently excreted. However, some researchers maintain that the levels found after handling thermal paper, around 20 nanograms per milliliter, are comparable to those that in epidemiological studies have been associated with health effects such as obesity, miscarriage, reduced libido, impaired sperm quality, and altered immune, liver, thyroid, and kidney function. These studies, though, are just associations and cannot prove a cause-and-effect relationship. For example, diet can influence both the amount of BPA ingested, since it is found in many canned foods, as well as the rate at which it is excreted in the urine. So a higher urinary level of BPA may just be a marker for a different diet or a different level of hygiene, both of which may account for the health effects rather than BPA.

Nevertheless, it seems advisable to avoid unnecessary exposure to BPA and the wearing of gloves by cashiers is a good idea, although nobody has studied what chemicals may be transferred to foods when handled by gloved cashiers. But there is enough concern about BPA on thermal paper to stimulate researchers to find alternatives, and some are already apparently available, since in studies not all such paper is found to contain BPA. One other problem is that when thermal paper is discarded and ends up in the waste stream, it can transfer BPA to other papers and plastics and end up in recycled materials from where it can again enter the body upon handling. Of course, we have to keep in mind that BPA is just one of the thousands of chemicals that find their way into our bloodstream through ingestion, inhalation, or skin exposure. Many of these, including a host of naturally occurring compounds found in soy, flax, cabbage, and licorice, as well as synthetics like PCBs, perfluoroalkyl substances, phthalates, and pesticides, have hormone-like effects. BPA is just one of many suspects when it comes to examining the effects of chemicals on health. But it is a suspect to keep an eye on.

SHINING A LIGHT ON THE IMPORTANCE OF DARKNESS

We may have been in the dark about light. Considering accumulating evidence, I have been trying a sleeping mask. That's to keep the blue light from the clock in my bedroom from launching an attack on my pineal gland. Alright, let's shed some light on what I'm talking about.

For most of human history, bright daylight alternated with nights of near-total darkness. Then electricity came along, primed to erase darkness. We now go to bed later and often sleep with night-lights that pierce the dark. In bed we watch TV, surf the web on laptops, and check our cell phones. It's all fun and convenient, but bathing in all this light may come at a cost. Especially if that light is blue.

Our bodies adhere to a twenty-four-hour biological clock, or circadian rhythm, that guides us when to sleep, rise, and even eat. This clock runs on melatonin, a hormone secreted by the pineal gland in the brain that communicates information about environmental lighting to various parts of the body. When little light enters the eye, the pineal begins to crank out melatonin, lulling us into sleep. As morning comes around and the sun makes an appearance, the production of melatonin winds down, signaling the body to rise and shine. Besides controlling sleep, melatonin triggers a host of biochemical activities such as reducing the nocturnal production of estrogen and testosterone. This is of concern because hormone fluctuations play a role in some cancers. Furthermore, melatonin itself has anticarcinogenic properties, suggesting that light can actually be a drug, amenable to abuse.

Over thirty years ago, researchers showed that exposing men to bright white light at 2 A.M. shut down the production of melatonin, just when it should have been at its peak. Then a report from the Centers for Disease Control and Prevention in Atlanta compared breast cancer rates in sighted and blind women and concluded that blind women were half as prone to develop the disease. A possible rationale was that their eyes could not detect light, making them resistant to fluctuations in melatonin and consequently other hormones as well. Then a study from Sweden backed up these observations, further noting that blind men were less likely to suffer from cancer, particularly of the prostate.

Animal experiments also show relevant results. Liver-cancer cells transplanted into rats grow more rapidly when the rodents are exposed to constant light than when kept in darkness. Even exposure to low-level light, such as that leaking through a doorway, results in tumor growth rate almost identical to that in animals exposed to constant bright light.

Other research noted that in animal tissues, a reduction in melatonin can suppress the immune system's ability to recognize and respond to emerging tumors. In view of such experiments, as well as the human epidemiological data, the International Agency for Research on Cancer (IARC) has concluded that people such as shift workers who regularly work in bright light at night could face a higher risk of cancer. As a result, IARC has placed night-shift work in category 2A, "probably carcinogenic to humans."

However, using the term "light" is like saying "flower." Not all light is the same. First, there is the question of brightness, usually measured in "lux." Then, as Newton's classic experiment demonstrated, passing sunlight through a prism separates it into a rainbow of colors, characterized by different wavelengths, ranging from long-wave red light to shorter-wave violet. We now know that blue light suppresses melatonin production more than other wavelengths. This is important information because fluorescent bulbs, light-emitting diode (LED) lights, and TV and computer screens emit more blue light than the older tungsten bulbs and sodium vapor lamps.

In light of the current trend of migrating towards LED technology in urban settings, and the consequent increased exposure to blue light, Spanish researchers used questionnaires to estimate indoor lighting, and images taken from the International Space Station to estimate outdoor blue light exposure. Then they compared people with prostate or breast cancer, with controls to see if greater exposure to blue light was linked with disease. Compared with those sleeping in total darkness, men who slept in "quite illuminated" bedrooms had a higher risk of prostate cancer whereas women had a slightly lower risk of breast cancer. When it came to outdoor lighting, blue light was positively associated with prostate cancer, and to a lesser extent, with breast cancer.

LED streetlights are more efficient and result in energy savings. They also make images sharper and increase visibility at night. So, what then do we do about the possible risk to health? Here, technology can come to the rescue. LED lights can be tuned to specific wavelengths, minimizing the range that can affect sleep cycles. Also, since such bulbs are more efficient, LED streetlights do not need to be as bright as older versions. At home, the best bet is to minimize exposure to screens before going to bed, admittedly a tough task. There is some evidence that wearing blue-blocker glasses when looking at screens in the evening can prevent melatonin suppression. Or you can don a sleep mask. A weighted one. It is claimed that the pressure, like a hug, brings about a state of peace and calm. We'll shed light on that notion at another time.

A BIBLICAL DYE

The ancient Phoenician city of Tyre on the Mediterranean coast was permeated by a putrid smell. It came from the processing of mounds of snails harvested from the ocean to yield the most famous dye of antiquity, Tyrian purple. As early as the seventeenth century BC, Phoenicians had discovered that extracts of three types of sea snails, *Murex brandaris*, *Thais haemastoma*, and *Murex trunculus*, were capable of yielding dyes ranging in shades from reddish to bluish purple.

Sea snails do not produce colored compounds to satisfy human vanity. They produce them to ward off predators. The very compounds that are prized for their ability to dye fabrics with stunning colors have flavors so bitter that predators learn to leave the snails alone. The three most noteworthy compounds the snails produce are dibromoindigo, monobromoindigo, and indigo, with the relative ratio of these determining the color of the dye that can be produced.

Phoenician dyers mastered the art of using the three types of snails to produce a variety of purple hues. Since huge numbers of snails were required to produce a small amount of dye, Tyrian purple was more expensive than gold! Given that only the very rich could afford purple fabrics, the wearing of such apparel became a status symbol. This was particularly the case in ancient Rome, with purple togas being restricted to the emperor and victorious generals. Julius Caesar wore a purple toga, and legend has it that Cleopatra's ship had sails dyed with Tyrian purple. After the fall of the western Roman Empire, the Eastern Empire, eventually known as the Byzantine Empire, continued to revere purple. Emperor Justinian I was routinely clad in purple and women of royal lineage gave birth in rooms decorated with purple fabric, giving rise to the expression "born into purple."

After about the seventh century, historical references to Tyrian purple faded and by all accounts the Middle Ages were characterized by drab apparel. It was in 1858 that interest in the ancient dye was rekindled when French zoologist Henri de Lacaze-Duthiers, on a trip to Spain, watched in amazement as a fisherman smeared his shirt with the slimy exudate of a snail, leaving a stain that at first was yellow but then turned to purple! His curiosity aroused, Lacaze-Duthiers rediscovered the three mollusks that were capable of producing purple-blue dyes. He also suggested that one of these, *Murex truculus*, was the source of the legendary blue dye referred to in the Old Testament.

"God said to Moses, speak to the children of Israel and say to them, that they shall make for themselves tsitsit on the corners of their garments throughout their generations. And they shall place upon the tsitsit of each corner a thread of tekhelet that they shall see and remember all of the commandments of God." Tsitsit are specially knotted ritual fringes, or tassels, on garments worn in antiquity by Israelites and today by observant Jews as a constant reminder to live according to God's laws. According to the Bible, one thread on each corner was to be colored with tekhelet, a blue dye. There is no mention of the source of the dye and the only information available comes from some esoteric references in the Talmud, a compilation of discussions and debates by Jewish scholars about the Torah, the Hebrew Bible. The Talmud, completed around the sixth century, suggests a snail known as hilazon as the source of tekhelet and describes the dye as a color similar to the sky and sea.

The problem is that there are no historical records of any snail that can produce a blue dye resembling the color of the sky. However, a chance discovery in 1980 by Israeli dye chemist Otto Elsner may have solved the mystery. While researching methods that may have been used by ancient dyers, he was working with extracts of *Murex trunculus*. Because of the smell involved, he worked near an open window. As expected, after extracting the snail's glands he produced a purple dye. But one day, brilliant sunshine came through the window, and when Elsner lifted his wad of dyed wool from the solution he was stunned to see it turn from purple to a brilliant blue!

As further research would show, energy from the sun's ultraviolet rays broke the bonds between carbon and bromine atoms in bromoindigo, yielding pure indigo, the same blue color available from the indigo plant. Ancient dyers working in the sunshine of the Middle East could well have noted this reaction and used it to produce tekhelet. Baruch Sterman, an Israeli scientist who is also a scholar of Jewish history, has taken an interest in producing the blue dye he believes is tekhelet from snails. He now gives demonstrations of the method at his workshop to visitors and turns out cotton fibers that once again observant Jews can use as a component of the fringes on their garments, reminding them of the ever-presence of God, and hopefully of the wonders of chemistry.

One day while Hercules was strolling along the shores of Phoenicia with a nymph he loved, named Tyrus, his dog, who was running beside them, came upon a *Murex trunculus*, its head protruding from its trumpet-like shell. The dog quickly devoured the shellfish and came away with a mouth stained brilliant purple. Enraptured by the tint, Tyrus claimed a robe of that same striking shade as the price Hercules would have to pay for her hand.

INTERMITTENT FASTING

The *New England Journal of Medicine* is widely regarded as perhaps the most prestigious medical journal in the world. It has an acceptance rate of 5 percent, meaning that only one in twenty articles submitted is judged by experts to be worthy of publication. That doesn't mean the rejected papers are not based on sound research; most eventually get published in lesser journals. But the *New England Journal of Medicine* looks for the cream of the crop. That's why I pay particular attention to papers published in this journal, such as a recent article on the "Effects of Intermittent Fasting on Health, Aging, and Disease." Even more so when the article is the work of Johns Hopkins University neuroscientist Dr. Mark Mattson, renowned globally for his research in the area of intermittent fasting.

Intermittent fasting does not mean cutting out a Snickers bar between meals. It refers to a systematic eating pattern that places emphasis not on *what* food should be eaten, but rather on *when* it should be consumed. Intermittent fasts fall into three general categories. In alternate-day fasting, days of very low calorie intake are alternated with days of regular eating. The 5:2 variety dictates eating normally on five days of the week but restricting calories to under 700 on two non-sequential days. In daily time-restricted feeding, all food is consumed in a six- to eight-hour window, essentially resulting in a sixteen- to eighteen-hour fast. Most people who engage in this version finish supper by about 7 P.M. and do not eat again until lunch the next day.

Of course, the question is why anyone would want to engage in any of these torturous regimes. Simply put, it is because there is accumulating evidence that calorie restriction provides benefits beyond the obvious weight loss. It has long been known that reduced food intake in animals results in an increased life span. The assumption has been that the benefits of reduced calorie intake are due to a reduction in the generation of free radicals as a consequence of metabolic processes. However, it appears that there is another factor involved. Typically, in experiments in which rodents are put on a low-calorie diet, they are given their daily allotment of food in one dose that they generally consume within a few hours. This means that essentially, they are on a twenty-hour fast. This results in metabolic switching, a term with which we have to become familiar to understand the benefits that are attributed to intermittent fasting.

The main source of energy for cells is glucose. During respiration, glucose serves as fuel, providing energy as it reacts with oxygen to yield carbon dioxide and water. It is this process that is also accompanied by the production of those troublesome free radicals. The main source of glucose is carbohydrates in the diet, and when these are severely restricted, as in fasting, the body switches to fats as an alternate fuel. But fats are not used directly; they are converted in the liver to ketone bodies that then are metabolized, yielding energy. This is commonly referred to as a state of "ketosis."

It turns out that these ketones are not just fuel for cells, but are also signaling molecules that regulate the expression and activity of various proteins and other biochemicals that influence health and aging. It seems that metabolic switching, which is a result of periods of fasting, is perceived by the body as a signal to go into survival mode since no food is coming in. Cells respond by improving control of blood pressure and blood sugar, producing more antioxidants, and curbing inflammation.

Most of the fasting studies that have produced promising results have involved animals, but some human trials are starting to emerge. Improvements in insulin sensitivity, verbal memory, resting heart rate, and cholesterol levels have been noted in short-term clinical trials. In rodents, experiments have shown reduced occurrence of spontaneous tumors with daily calorie restriction or alternate-day fasting. Suppression of the growth of induced tumors has also been observed. Furthermore, the animals show increased sensitivity to radiation and chemotherapy. Stimulated by these observations, a number of human trials examining the effect of intermittent fasting on breast, ovarian, prostate, endometrial, colorectal, and brain tumors are underway. Pilot studies are also examining the possible benefits of intermittent fasting in multiple sclerosis, rheumatoid arthritis, surgical outcomes, and athletic performance.

Obviously, we have to temper this discussion with the all-too-often-stated disclaimer that more research is needed. But it is forthcoming. Dr. Valter Longo at the University of Southern California has some interesting results with his Fasting Mimicking Diet. For five days, people consume only special prepackaged foods that provide 1,000 calories the first day and 725 the other days and are said to have a unique combination of nutrients that trick the body into thinking it is fasting. Repeating the cycle monthly for three months has resulted in weight loss as well as a drop in blood sugar and cholesterol. But going hungry for five days is challenging and the meals are expensive.

It is always meaningful to ask experts what change they have made in their life as a result of their research. Dr. Mattson says he eats within a six-hour window every day. And that's from the horse's mouth.

RECYCLING POLYSTYRENE

As you may know, I'm into both chemistry and magic. I often make the point that chemistry can appear to be magical until an explanation makes the magic vanish and replaces it with science. A typical example is a demonstration I have often performed in which a polystyrene coffee cup is slowly inserted into a beaker of acetone and appears to disappear. Of course, one of the fundamental laws of chemistry is that matter cannot be created or destroyed, it can only be changed from one form to another. In this case, the polystyrene doesn't vanish, it just dissolves in the acetone, much like sugar dissolves in water.

As the term implies, polystyrene is made by linking styrene molecules together into a long chain. Styrene in turn is made from benzene and ethylene, derived from petroleum. If during the polymerization of styrene, a mixture of gases such as pentane and air is injected, bubbles form within the plastic and the resulting product is known as foamed or expanded polystyrene. This can then be molded into coffee cups as well as a host of other products ranging from food packaging and insulation panels to yogurt containers and packing peanuts. During processing, pentane evaporates and the bubbles in the finished plastic contain only air. The tiny bubbles lower the density of the material, making foamed polystyrene very light. At one time, chlorofluorocarbons (CFCs) were used as the blowing agent with the belief that these were inert substances with no environmental consequences. That assumption turned out to be incorrect when it was discovered that CFCs were instrumental in the destruction of the ozone layer. This resulted in the switch to air and pentane as the blowing agent.

Whether a substance dissolves in a solvent is determined by the ease with which molecules of the solvent can insert themselves in between molecules of the solute. In the case of polystyrene, acetone molecules can readily force themselves in between the polystyrene molecules, causing polystyrene to dissolve. In the case of a foamed polystyrene cup, the dissolution is very quick because the plastic content is actually quite small; most of the volume is taken up by air. To demonstrate that the coffee cup has not actually vanished, the acetone can be evaporated, leaving behind a small blob of polystyrene.

Polystyrene, whether foamed or not, is notoriously difficult and expensive to recycle. Because of the volume that foamed polystyrene products take up, economical transportation is a problem. Also, when food containers are involved, they are often contaminated with the food that has been stored in them. As a consequence, polystyrene often ends up in landfill or being incinerated, neither of which is a desirable option. With the increasing concern about plastic pollution of our environment, a great deal of effort is being directed towards recycling. Now, a Quebec company, Polystyvert, may have come up with a process to recycle polystyrene into pellets that can then be remolded into new products. And the technology is essentially based on the vanishing coffee cup magic trick.

The idea is to first use a solvent that will dissolve polystyrene at a collection site and then transport the solution to the recycling facility.

This reduces transportation costs by eliminating the need to transport voluminous foamed plastic. Obviously, the first requirement is to find a suitable solvent that is economical and can be used safely. Para-cymene, a naturally occurring compound found in wood, and easily available as a byproduct of the pulp and paper industry, fits the bill. It readily dissolves polystyrene but will not dissolve other plastics such as polyester, polyethylene, or polyvinyl chloride (PVC). This is important because the bane of recycling is the separation of different plastics, since each requires different recycling technologies.

Once the p-cymene solution is transported to the recycling facility, it is filtered to eliminate insoluble impurities. Next, pentane is added to the filtrate, and since polystyrene is not soluble in this solvent, it precipitates out of solution. The mixture of solvent and precipitate is now heated, driving off the pentane and p-cymene, which are collected and recycled. During this distillation process, various additives such as flame retardants in the original plastic are also driven off. Since pentane and p-cymene have low boiling points, the distillation process does not require much energy. Polystyvert claims that its recycled polystyrene is 40 percent cheaper than virgin polystyrene made from fossil fuel and that its process releases far less greenhouse gases than making polystyrene from scratch. Furthermore, the recycled polystyrene is so pure that it can be used for food-contact applications, and the company has already filed for approval of such use with the U.S. Food and Drug Administration.

Whether this novel recycling process will progress beyond the pilot-plant scale remains to be seen, since there are other competing technologies. For example, researchers at the University of Minnesota have shown that with the use of a special platinum or rhodium catalyst, polystyrene can be decomposed into styrene, which can be collected and repolymerized. But the process does not solve the problem of transportation costs, and the catalyst is expensive.

For me, there has already been a benefit from learning about Polysyvert's technology. Next time I perform the vanishing coffee cup demo, I'll be able to expand on the story and talk about how chemists may have found an interesting "solution" to an environmental problem.

SUPERABSORBENT POLYMERS REALLY SUCK IT UP

On March 18, 1965, Soviet cosmonaut Alexei Leonov climbed through the hatch of his Voskhod 2 capsule and became the first man to walk in space. He later wrote that it felt "like a seagull with its wings outstretched, soaring high above the Earth." Since the exhilarating experience only lasted ten minutes, he did not have to worry about having to answer nature's call. It was quite a different case for NASA astronauts Jim Voss and Susan Helms in 2001 when they spent almost nine hours outside working on the space station. There would have been no way to climb back inside, remove the cumbersome space suit, and engage the vacuum hose–equipped funnel that suctions urine away, eventually to be converted into potable water. Fortunately, there was no need for a frantic trip to the space toilet, thanks to the special undergarment the astronauts wore. It contained sodium polyacrylate, a superabsorbent polymer capable of retaining at least thirty times its weight in urine!

By the 1930s, chemists had learned how to link certain types of small molecules into long chains called polymers. The properties of these giant molecules depended on the small molecules, or monomers, from which they were constructed, as well as on whether the chains were joined to each other by "cross links." In 1938, German chemist Werner Kern noted that a polymer he had synthesized from acrylic acid and a cross-linking agent, divinylbenzene, had a remarkable property. Particles of polyacrylic acid, as they called it, had an amazing ability to absorb water, swelling to many times their initial size. This didn't get much attention until the 1960s, when the U.S. Department of Agriculture developed an interest in water conservation in soils. Researchers synthesized a variety of polyacrylates and found that some, sodium polyacrylate for example, were capable of absorbing hundreds of times their weight in water.

At the time, the agriculture industry did not embrace the potential of these substances, but chemists at the Dow Chemical Company had an idea. These superabsorbent polymers seemed ideal for use in diapers and feminine personal care products. Although they worked well, sales languished until Japanese companies got into the market in the 1980s. Sales boomed in Japan, boosting American interest. By the end of the decade, superabsorbent polymers had become a giant industry with applications extending way beyond diapers and personal care products.

Leaking water is a huge problem in the construction industry. A superabsorbent powder can be blended with rubber to form a composite that can be used as mortar between cement blocks. On exposure to water, the composite swells and prevents water from penetrating. This technology was especially useful in the construction of the Chunnel, the tunnel between England and France. Underground cables can also be protected from water damage by being wrapped with a tape containing superabsorbent polymers to intercept water before it can cause harm. These polymers can also be used in the pads found in meat and poultry packages, where they absorb fluids that would otherwise be a breeding ground for bacteria. And the original idea of using these chemicals in agriculture and horticulture has been revived.

In this case, potassium polyacrylate is used instead of sodium polyacrylate because excess sodium can damage plants, while potassium actually acts as a fertilizer. Coating seeds with the substance can lead to quicker emergence of the plant, and blending into soil can reduce the amount of water needed for irrigation since water is retained by the polymer and is released as needed. It can even be used at home for potted plants, which can then be watered less often. No need to ask the neighbors to water your plants when on vacation.

How do superabsorbent polymers work? The key is osmosis, a

process by which molecules of a solvent tend to pass through a semi-permeable membrane from a less concentrated solution of a solute into a more concentrated one, trending towards equalizing the concentrations on each side of the membrane. In this case, the cross-linked polyacrylate molecules act as a membrane, the solvent is water, and the sodium or potassium ions embedded in the cross-linked matrix constitute the solute. Since there are more ions inside the matrix than outside, water flows in, swelling the polymer.

A classic magic trick is to place a small amount of "super slurper" powder into the bottom of an opaque cup, pour in water, and then turn the vessel upside down, showing that the water has disappeared. Of course, it is still in there in the form of a gel that will not fall out. Now you can add a "magic chemical," which is actually a bit of salt, and then pour out the water that has wondrously reappeared. The addition of the salt means that now there is a higher concentration of sodium outside the polymer than inside, so water will flow out of the matrix.

This can be useful knowledge if someone has accidentally put a diaper down the toilet, where it can absorb water and swell, causing a blockage. Just sprinkling in some salt can prevent an expensive visit by a plumber. I don't know if this technique is used to recover water from the astronauts' superabsorbent garments for recycling into drinking water, but it could be.

ARSENIC, THE KING OF POISONS, POISON OF KINGS

Ask someone to name a poison and chances are they will come up with arsenic. Even though compounds of mercury, beryllium, cadmium, thallium, and polonium are more toxic, as is botulin produced by *Clostridium botulinum* bacteria or ricin found in the castor bean plant, arsenic gets more publicity. You may recall how lonely old men were knocked off with poison-laced elderberry wine in the classic 1944 film *Arsenic and Old Lace*, or remember having read one of Agatha Christie's many stories in which arsenic was the villain. History buffs may harken back to accounts of the Borgias eradicating rivals with arsenic or the exploits of Giulia Toffana, the seventeenth-century Italian professional poisoner, who specialized in creating wealthy widows. Then there is the popular theory that Napoleon was either purposely poisoned with arsenic by his British captors or succumbed to vapors emanating from wallpaper that had been colored with arsenic compounds.

It is hard to determine exactly when arsenic was first recognized as being toxic, but that discovery likely dates back to the first smelting of metals from their ores some 8,000 years ago. Arsenic-containing minerals such as orpiment or realgar, which are sulfides of arsenic, are sometimes found together with other metal ores. When heated in the presence of oxygen, arsenic sulfides convert to arsenic trioxide, which is then emitted as a white smoke. The toxic nature of the smoke would have become obvious to early metallurgists and it would not have been difficult to trace the toxicity back to the arsenic minerals.

In any case, we know that the toxic properties of arsenic were described as early as the fourth century BC by Hippocrates, and that in AD 55, Dioscorides, a physician in the court of the Roman Emperor Nero, recorded how Nero poisoned his stepbrother Britannicus with arsenic to secure his position as Roman emperor. Curiously, Hippocrates also used arsenic sulfides to treat ulcers and abscesses, and Dioscorides used them as a depilatory. Arsenic compounds were also mainstays in ancient Chinese and Ayurvedic medicine as treatments for numerous conditions ranging from dermatitis to indigestion. There is no evidence that they were useful for any of these but could well have been responsible for side effects such as skin pigmentation or anemia.

By the Middle Ages, arsenic had gained a reputation as the "king of poisons" and the "poison of kings." The Medicis and the Borgias, both papal families, rode to power on the coattails of arsenic, and Catherine de' Medici introduced Italian poisoning methods to France when she married the future King Henry II. It seems the king was aware of his intended's talents because he insisted that a unicorn horn be part of the dowry. It was believed at the time that a unicorn horn would offer protection against poisons. Quite a stretch since unicorns are mythical creatures. But rhinos and narwhals are real, and their horns served to please the gullible. Needless to say, they would have offered no protection against arsenic.

There is no question that royalty were worried about being poisoned, and human "tasters" were often employed as poison detectors. Kings in the ancient Korean kingdom of Baekje had a different idea. They took to using silver chopsticks to detect arsenic in their food! Apparently they had become aware of the fact that when silver comes into contact with certain arsenical minerals, it becomes discolored. Today, we would say it tarnishes. The chemistry here is interesting. Silver reacts with sulfides to form black silver sulfide, the bane of silverware lovers. Any sulfide will do, including the traces of hydrogen sulfide normally present in air, or arsenic sulfide that may be present in food. Interestingly, the toxicity of arsenic is also due to its reaction with sulfur compounds, in this case with sulfur-containing enzymes that are critical to body functions.

Of course the Korean kings didn't know the chemistry involved in the tarnishing of silver, but they were aware that silver would discolor when exposed to powders made from minerals that at the time were recognized as being poisonous. Hence the introduction of silver chopsticks. It is unlikely that this would have been effective because even if there were a significant amount of arsenic sulfide in the food, the reaction takes quite some time to develop. Also, there are many naturally occurring sulfides in food. Cabbage, eggs, mustard onions, and garlic are examples of foods that could lead to the tarnishing of silver. Furthermore, other forms of arsenic, such as arsenic trioxide, would not tarnish silver and would not have been detected.

It is interesting to note that of all the Asian countries that have a tradition of using chopsticks, only the Koreans use the metal variety. Some historians believe this practice dates back to the kings' use of silver chopsticks. People yearned to eat like the king, but could not afford silver and resorted to chopsticks made of cheaper metals. Maybe so. Maybe not.

ARSENIC IN FACT AND FICTION

You would not think that former American president Zachary Taylor and English novelist Jane Austen would have anything in common, but they do. Both have been the subject of theories that their deaths, commonly attributed to natural causes, were actually the result of poisoning by arsenic.

Zachary Taylor died in 1850, just sixteen months after being elected president. On a hot July fourth, he had attended the groundbreaking ceremony for the Washington Monument and then headed back to the White House, where he consumed a large bowl of cherries and a glass of cold milk. By the evening he began to feel unwell, and by the next morning, he was experiencing bloody diarrhea. Diagnosed with acute gastroenteritis or possibly cholera, he was treated with mercury chloride (calomel), ipecac, and opium, according to best practices at the time. The calomel would have been useless, but opium can curtail diarrhea. Ipecac induces vomiting, which was believed to rid the body of whatever was invading it. There is also a good chance that the president would have been bled, a common treatment for all ailments at the time. If anything, these treatments hastened his death, which occurred just four days later.

Posthumous diagnosis is notoriously treacherous and is usually based on a lot of conjecture. Judging by the rather sudden onset of symptoms, some sort of food poisoning, either by bacteria such as *Salmonella* in the cherries or in the unpasteurized milk, seems a possibility. However, author Clara Rising, after doing research on the president for a book she was writing about his life, introduced another theory. She claims that Taylor's symptoms are consistent with arsenic poisoning and that the president had plenty of enemies who would have profited from his death. Security in those days in the White House was not tight, and tainting some food with arsenic would not have been difficult. Although Taylor himself was a slave owner, he proposed that any new territories that joined the U.S. would be free states, a proposal that angered many Southerners.

Rising made a case for her views in *The Taylor File: The Mysterious Death of a President* and managed to obtain approval from Taylor's descendants for an exhumation and forensic autopsy. If arsenic were found, it would pretty well be an indication of poisoning because the body had not been embalmed, a process that in those days used arsenic. The president had been in a mausoleum, not buried in the ground, meaning no arsenic contamination from the soil. The autopsy found no arsenic and came to the conclusion that the cause of death was cholera or acute gastroenteritis, as Washington had open sewers and his food or drink may have been contaminated. But that did not satisfy Rising. She suggested that he may have been poisoned with cyanide or with mushrooms, neither of which would have left a trace after 140 years. If Rising's allegations were correct, Zachary Taylor would have been the first U.S. president assassinated. However, history bestows that dubious honor on Abraham Lincoln.

Jane Austen was one of the most famous novelists of the nineteenth century, rivaling Charles Dickens in popularity. Her *Pride and Prejudice* and *Sense and Sensibility* were widely read and there was great sadness when Austen died at the young age of forty-one. Cause of death has been attributed either to underactive adrenal glands (Addison's disease), the autoimmune disease lupus, or to some form of cancer. As in the case of Zachary Taylor, a novelist has now suggested that arsenic killed Austen, albeit probably not by deliberate poisoning. Lindsay Ashford came across some of Jane Austen's letters and was alerted by the sentence, "I am considerably better now and am recovering my looks a little, which have been bad enough, black and white and every wrong color." Since Ashford was a crime writer and knew about poisons, she thought that the changes in pigmentation that Austen described smacked of arsenic poisoning.

In the eighteenth and nineteenth centuries, various forms of arsenic were used as medicine. A 1 percent solution of potassium arsenite, known as Fowler's solution, was a common treatment for almost any ailment, despite being useless. Although there is no record of Austen having used this "remedy," Ashford suggests that it is very likely since it was used for rheumatism, something that Austen complained of in her letters.

Ashford receives some support for the theory of arsenic poisoning from Sandra Tuppen, a curator at the British Library. She bases her argument on three eyeglasses of different strengths that belonged to Austen and are now displayed in the British Library. The argument is that Austen needed to get stronger and stronger glasses because she was suffering from cataracts, and guess what, arsenic can cause cataracts. Seems like quite a stretch since there are many reasons why people's eyesight can change. And unlike Zachary Taylor's case, we will never find out whether her body contains arsenic because Jane Austen will never be exhumed. She lies in peace under the floor of Winchester Cathedral.

WIG POWER

William Pitt the Younger became prime minister of Britain in 1783 at the tender age of twenty-four. The term "Younger" distinguished him from his father, William Pitt the Elder, who had also served as prime minister. During the younger Pitt's tenure, Britain was embroiled in a war with Napoleon that placed a heavy burden on the treasury. To raise money, Pitt introduced a number of taxes, including the Duty on Hair Powder Act of 1795. Anyone wishing to purchase hair powder had to first acquire a certificate from a post office upon payment of an annual fee. Since hair powder was a popular commodity among the fashionable nobility, the tax brought in significant funds.

The story of the appeal of hair powder starts in the sixteenth century when many nobles were plagued with syphilis, a sexually transmitted disease that can result in open sores and hair loss. Long hair was in style at the time and hair loss was often a giveaway of having contracted syphilis. Wigs masked the situation but introduced another problem. They were usually made of the hair of goats or horses that attracted lice and also had a smell. Finely powdered starch, or "hair powder," was found to deter lice since its slippery nature prevented the bugs from climbing the hair shaft to get to their food supply — namely, the scalp. Often, the starch was scented with orange, jasmine, or lavender essence to mask any unpleasant smells.

Wigs as a fashion statement started with Louis XIV, the Sun King, who was also a king of fashion. He introduced red shoes as well as long hair that courtiers took to imitating with wigs. The wearing of wigs was not novel. Much earlier the ancient Egyptians had fabricated wigs from human hair, wool, or flax fibers, and the Romans used the blond hair of northern captives. Roman ladies of the night were recognizable by the blond wigs they sported. Louis's flowing locks had defined his appearance, and when he started to lose his hair, he began to resort to elaborate wigs. At one point the king had forty wigmakers and owned some 1,000 wigs made from human hair!

The towering audacious wig hairdos required a great deal of maintenance. Heated clay rollers produced the required curls and a "pomade" kept them in shape. That term derives from the Latin *pomum* for apple — since historically, mashed apples were combined with animal fat to make pomades. The sugar present in the apples enhanced their stickiness, which was welcome not only to maintain the shape of the locks but also to help bind the wig powder to the hair. The powder was sometimes colored with minerals, but pure white from finely macerated flour was the most popular. The pomade and the powder added significant weight to the wig, and the bewigged had to learn how to balance the elaborate beehives as they walked.

Wigs made it across the English Channel when Charles II returned to England from France after having been exiled by Oliver Cromwell. He had taken a shine to Louis's wigs, and when he was restored as king, the English nobility followed in his footsteps. By the end of Charles's reign in 1685, wigs had become part of barristers' and judges' wardrobe in the court, a tradition maintained to this day. Wigs made from the hair of white horses were the most prized, but as they tended to yellow with age, they required the use of white powder.

Gentlemen who did not favor wigs, George Washington for example, had their own hair pomaded and powdered to look as if they were wearing a wig. The president's white, pomaded locks can be readily seen in Gilbert Stuart's famous 1796 portrait, painted just a year after the Duty on Hair Powder Act had been introduced in Britain. Of course, since America had broken from Britain, the Act did not apply in the newly formed United States. It is hard to know just how much revenue the Act contributed to the British coffers, but Prime Minister William Pitt apparently did his part. A portrait shows his hair coiffed with white powder.

Wigs as a style statement for men have disappeared except in British courts and some former colonies like Hong Kong, where they are still worn by lawyers and judges. The argument is that they provide a certain anonymity, resulting in the court's attention being focused not on the person but the law. Designs of the wigs are different for lawyers and judges, with judges' wigs being more ornate and longer. Made from white horsehair, such wigs can cost thousands of dollars. I don't know whether lawyers or judges still powder their wigs, but wig powders still exist, although they are usually called setting powders. And there is no extra duty on them.

PLANTS AND HOSPITALS

The garbage heap of science is filled with ideas and practices that were at one time thought to be on the mark. Lobotomies for mental illness, lying immobilized after cataract surgery, and bed rest for heart attacks were once common practices. And if you were hospitalized, plants that visitors brought to introduce some cheeriness were routinely removed from the room at night. This was based on the erroneous belief that the plants use up oxygen at night and deplete the oxygen in the air that is available for patients to breathe.

While it is true that plants use up oxygen at night through the process of respiration, the amount is insignificant in comparison to the amount of oxygen contained in air. A visitor breathing in a patient's room uses up much more oxygen. Over a twenty-four-hour period, plants produce more oxygen through photosynthesis than they use up through respiration.

During the day plants carry out two processes. They photosynthesize and they respire. Photosynthesis is perhaps the most critical chemical reaction in the world because without it there would be no life. It is what allows plants to grow, and all life either directly or indirectly depends on plants. Through photosynthesis, plants absorb carbon dioxide from the air and combine it with water to produce glucose, which then serves as the starting material for the biosynthesis of carbohydrates, proteins, fats, vitamins, and the diverse other molecules that make up a plant. As the term implies, "photosynthesis" is dependent on light. ("Photo" derives from the Greek for "light.") It is also dependent on chlorophyll, the green pigment responsible for the absorption of the light, which provides the energy needed for photosynthesis. It is often said that chlorophyll, from the Greek for "green" and "leaf," is the most important chemical in the world because without it plants cannot grow. And without plants neither can we.

However, glucose is more than just a starting material for the synthesis of other compounds. Through the process of respiration, which is the reverse of photosynthesis, it serves as the fuel needed to produce the energy that powers the plant's growth. Basically the plant "burns" glucose, reacting it with oxygen to yield carbon dioxide and water with the simultaneous release of energy.

When light is available, both photosynthesis and respiration occur, but at night, in the absence of light, there is no photosynthesis. Respiration, however, continues, using up oxygen. But the amount used does not make a dent in the amount of oxygen present in air. There is no need to remove plants at night from the room.

There could be a minor concern about cut flowers in a vase, since the water could, in theory, be a breeding ground for bacteria. However, no infections in hospitals have been linked to flowers. And there is no question that flowers can brighten a patient's mood. In fact, several studies have shown that after surgery, patients who have plants or flowers in their room require fewer pain killers, have fewer complaints, and recover faster. The greenery does not necessarily have to be in the room. Patients with windows that look out onto trees do better than when the windows face a concrete jungle. Interestingly, pictures of plants in the room have the same effect as actual plants.

There is another reason people think plants are a good idea. Purification of the air! Unfortunately, that just isn't the case. Back in the 1980s, a NASA researcher did carry out some studies to investigate methods of removing volatile organic compounds such as formaldehyde or benzene from the air. The concern was that in the enclosed environment of a space capsule, some of the plastics used might outgas some troublesome compounds. He did show that some plants did absorb some of these from the air. However, the catch is that those studies were done in a sealed room the size of a space capsule with no air circulation. When similar studies were carried out under realistic conditions such as would be encountered in a home, there was no significant removal of air pollutants. The number of plants needed to purify the air to any significant extent would have to fill the room, leaving no space for any occupants.

If there is a concern about volatile organic compounds, a bag of activated charcoal would rise to the occasion. These work really well for any odor as well as for odorless compounds released by molds. Charcoal has an amazing ability to bind molecules to its surface, and activated carbon has a huge surface area because each grain is permeated by microscopic tunnels. Once the charcoal becomes saturated with the substances it has adsorbed, it ceases to be active, but the good news is that it can be reactivated just by placing it outdoors in the sun for a few hours. The adsorbed materials will be released and the charcoal is then ready to pick up another cargo.

CHALK IT UP

My first encounter with calcium sulfate was when I broke an ankle, way, way back in grade eight. I had no idea of the chemistry going on when the gauze bandages impregnated with some white stuff were wrapped around my ankle, hardening into a cast. Later, I would learn that the "white stuff" was plaster of paris, or in chemical parlance, calcium sulfate hemihydrate. When moistened, plaster of paris is easily shaped, but hardens into a lightweight mass as it absorbs water and forms calcium sulfate dihydrate, better known as gypsum. Plaster of paris is actually anhydrous calcium sulfate, prepared by heating gypsum to drive off water. It derives its name from the discovery of large deposits of gypsum in the Montmartre region of Paris in the eighteenth century.

I bumped into calcium sulfate again in the undergraduate chemistry lab where we used the anhydrous version as a desiccant to dry chemicals that we had synthesized by placing them in a desiccator, a glass container with a tight-fitting lid and a desiccant at the bottom to absorb moisture.

Once I started teaching, I would have yet another meeting with calcium sulfate. That would be on the blackboard. Organic chemistry involves drawing numerous molecular structures, and back in the days before PowerPoint, the blackboard was the way to go. I had been inspired to work on my blackboard technique after attending a lecture by Robert Woodward, a Harvard professor and Nobel Laureate who was renowned for his syntheses of extremely complex molecules, as well as for his blackboard skills. I practiced and practiced, but whereas Woodward's chalk had seemed to glide over the board, mine sputtered with stops and starts. Furthermore, his lines were broad and impactive, mine were squiggly. Then one day, lightning struck.

At my daughter's birthday party, she got a present, a box of sidewalk chalk. I wasn't thrilled by the potential consequences of this present and said that I would hang on to it until she was older. But when I picked up a piece of the chalk, it didn't feel like the one that I had been using. It was much softer, and thicker. Could my blackboard problems have been due to using the wrong type of chalk? A little investigation revealed that chalk could be made of calcium carbonate or calcium sulfate. It wasn't hard to distinguish between the two. A few drops of hydrochloric acid applied to my chalk quickly resulted in the formation of bubbles of carbon dioxide, indicating that it was made of calcium carbonate, I quickly purchased some calcium sulfate blackboard chalk.

What a difference! It practically floated across the board, leaving behind nice broad lines. I never went back to the carbonate again! However, there was a downside to the change. When I had lectured about antacids, I'd always described how some of the most common ones were composed of calcium carbonate, since as a base, it was capable of neutralizing stomach acid. For emphasis, I would take a bite out of my chalk. I couldn't do that with calcium sulfate since it does not neutralize acids. In the coming years, many students would eventually tell me that the lecture they remembered most was when their crazy professor ate chalk!

I was happy with my calcium sulfate chalk, although it did break easily and produced more dust than I liked. Only recently did I come to learn that I could have done better. I could have gone for the Rolls-Royce of chalks, the one that mathematicians have described as the "nectar of the gods," one that melts like butter when you start writing. That is the Hagoromo Fulltouch Chalk, made by Hagoromo Bungu, a

Japanese company. It seems this company had a process for formulating calcium carbonate into a soft chalk that was then coated with a waxy layer, preventing it from breaking or giving off too much dust. Alas, the company stopped production in 2015, saying that blackboards had gone out of fashion, giving way to whiteboards and "smart" boards. At this devastating news, mathematicians started to buy up all available supplies and hoarded the chalk. Hagoromo's equipment was eventually sold to a Korean company that now markets the "umami of chalk."

However, some claim that it is not as good as the original, a very limited supply of which is available from Amazon at $849 for seventy-two pieces. That's crazy, but the new version is available for $49.50. Of course, I had to get it. It is not quite as good as my old calcium sulfate chalk, but it allowed me to eat chalk once more, although I'm not that fond of the waxy coating. But then to my surprise, I discovered that a number of companies make "edible chalk" sourced from deposits in the Ukraine. Got some. As it turns out, it also performs well on the blackboard. Life is good.

OMEGA-3 QUESTIONS

I do get some curious questions. "Are there any fish parts used in Omega watches?" It wasn't difficult to guess that the question was prompted by the association of "Omega" with fish since the omega-3 fats found in fish oil have received a great deal of media attention due to their supposed health benefits. Since omega is the last letter in the Greek alphabet, the term, in this case, is used to denote "the end" of the chain of carbon atoms that make up a fatty acid. The number 3 refers to the presence of a carbon-carbon double bond on the third carbon from the end. No such fats, or any other fish parts, are used in Omega watches. The name likely was coined to suggest that as far as watches go, this was the ultimate, "the end."

I don't have an Omega watch, but I do have an interest in omega-3 fats since I have a fish allergy, meaning that fish oils are not part of my diet. So, I have wondered about taking fish oil supplements, sourced not from fish but from algae, which after all is where fish get them from. The scientific literature on omega-3 supplements is vast, and as is so often the case, can be cherry-picked to show either that they have great health benefits or that they are useless.

The stage for the fish oil controversy was set in 1971 by Danish researchers Hans Olaf Bang and Jørn Dyerberg, who reported in *The Lancet*, a prime medical journal, their finding of low cholesterol and low triglyceride levels in the blood of Inuit peoples living in Greenland. In the study, they used the term "Eskimo," mentioning that the name derives from a native word meaning "people eating raw meat." That etymology is controversial, and "Eskimo" is now considered to be offensive. The preferred term is Inuit.

Finding low levels of cholesterol and triglycerides was surprising given that the Inuit diet was based mostly on animal products, which other studies had linked to high blood levels of the same substances. Bang and Dyerberg suggested that the difference was due to the type of animal fat consumed, which in the case of the Inuit came mostly from fish, seals, and whales. These fats fall into the polyunsaturated category and have different biochemical effects than the saturated fats in land-based animals. The authors of *The Lancet* paper also noted that the heart disease rate among the Inuit was much lower than among the general Danish population and hypothesized that this may be due to some protective effect offered by the marine fats.

Correctly, they clearly pointed out that such an observational study was not capable of proving a cause-and-effect relationship since other lifestyle factors may be involved. Perhaps it was the activity level of the Inuit that was protective, or maybe it wasn't what they were eating, but rather what they were not eating, namely the hamburgers, steaks, pizzas, and processed staples of a Western diet. Genetics did not appear to play a role. The Inuit living in Denmark had blood chemistries comparable to the rest of the Danish population.

Some have criticized Bang and Dyerberg, saying that their claim of fish fats offering protection against heart disease is not justified since they based their conclusion on unreliable reporting of the incidence of heart disease in Greenland. But the researchers never claimed that they had proven any such link. They just suggested that their study could be a springboard for randomized controlled trials using omega-3 supplements.

Indeed, there have been a number of such trials, the best of which were evaluated together in a meta-analysis, essentially a study of studies, published in 2018 in the *Journal of the American Medical Association*. This included ten randomized trials that involved at least five hundred participants and a treatment duration of at least a year, with various doses of supplements. The authors concluded that "omega-3 fatty acids had no significant association with fatal or nonfatal coronary heart disease or any major vascular events." Case closed? Nope. It never is.

In 2019, the results of a highly anticipated randomized, placebo-controlled trial involving over 25,000 healthy participants taking 1 gram of omega-3 supplements a day for five years were published in the *New England Journal of Medicine*. While supplement takers did not have a lower incidence of total cardiovascular events, there was a reduced incidence of heart attacks, especially among people who do not eat fish. Another paper published in the same journal reported on a study in which subjects who already had cardiovascular disease or diabetes were supplemented with 4 grams of eicosapentaenoic acid (EPA), a specific omega-3 fat. They had a 25 percent reduced risk of cardiac events!

Frankly, given that studies have used different doses of supplements, different compositions, and had different subject profiles, it is not possible to come to a firm conclusion about the potential benefits of supplements, although the evidence seems to be slightly in favor. But which one and how much?

There is better data on the Omega watch. It was worn by President Kennedy at his inauguration, James Bond wore it in several films, and it was selected by NASA astronauts to be the first watch to go to the moon! And that is no fish tale.

THE SPY WHO WENT OUT INTO THE COLD

Vladimir Putin strode to the podium in 2007 and announced that George Koval would be honored with the prestigious Gold Star as a Hero of the Russian Federation. The proclamation was especially noteworthy given that Koval, who had passed away the year before at age 92, was an American! What does it take for an American to become a Russian hero? Providing critical information needed to construct an atomic bomb will do it. George Koval, you see, was a Soviet spy. "Mr. Koval," President Putin stated, "operated under the pseudonym Delmar and provided information that helped speed up the time it took for the Soviet Union to develop an atomic bomb of its own."

Koval was born to Russian immigrant parents in Iowa, but as an adult spent time in the Soviet Union, where he was recruited by military intelligence and trained as a spy before returning to the U.S. in 1940. There he joined the army and by 1944 managed to secure a position as a "health physics officer" working at the Oak Ridge Laboratories in Tennessee, where as part of the Manhattan Project researchers were bombarding the nuclei of plutonium and uranium atoms with neutrons. The energy released when these nuclei were split into smaller fragments, or fissioned, was the key to developing the atom bomb. Koval began reporting on these activities to his handler; however his opportunity to make a real contribution to the Soviet nuclear program came in 1944 when he was transferred to a top-secret lab in Dayton, Ohio.

It was there that critical work was being pursued to develop an initiator for fission reactions. Once a fission reaction begins, it generates neutrons that then trigger a chain reaction, but a source of neutrons is needed to initiate fission. These can be produced by bombarding beryllium with alpha particles emitted by polonium-210, a naturally radioactive isotope. Alpha particles are composed of two protons and two neutrons and can act like minuscule bullets, in this case knocking out neutrons when they strike the nuclei of beryllium atoms. The problem was to secure a supply of polonium, a very rare element.

Back in 1898, Marie and Pierre Curie had processed tons of a uranium ore known as pitchblende to finally isolate a few micrograms of a residue that had much greater radioactivity than could be accounted for by uranium. They surmised that the residue must contain a novel element! Marie, who incidentally coined the term radioactive, suggested that the newly discovered element be named polonium, after her country of birth. To produce an initiator for the bomb, a more efficient way of procuring polonium had to be found. This was the task to be undertaken by the Dayton Project.

In 1942, Brigadier General Leslie Groves Jr., director of the Manhattan Project, had approached the Monsanto chemical company to oversee the purification process needed to produce plutonium. Charles Allen Thomas, who later became president of Monsanto, was put in charge of the project. When the need for a polonium initiator became clear, he was also asked to find an efficient method of production. The research was carried out at Monsanto's labs in Dayton, with a focus on isolating the desired polonium from the lead oxide ore in which it occurs naturally, albeit in very small amounts.

As chemists were struggling to isolate polonium from the lead ore, a more ready supply became available. The Manhattan Project had developed a nuclear reactor capable of bombarding bismuth with neutrons to produce the desired isotope of polonium. Now the task was to separate the polonium from residual bismuth. This was soon mastered, allowing enough polonium to be produced for both the uranium bomb that was dropped on Hiroshima and the plutonium bomb that annihilated Nagasaki. It was the production method of polonium that Koval transmitted to the Soviets, resulting in their developing a nuclear bomb much sooner than the Americans had expected.

The alpha particles emitted by polonium-210 can also target human tissues, causing severe damage. Indeed, polonium is one of the most toxic substances known. Ingesting or inhaling just a few micrograms can be lethal, as evidenced by the celebrated poisoning in 2006 of former Soviet secret service agent Alexander Litvinenko in London. Litvinenko had sought asylum in Britain after being charged with exceeding the authority of his position in Russia. He had specialized in tackling organized crime and had run afoul of the authorities with his claim that the Russian government was involved in state-sponsored terrorism. In England, he had authored two books that were highly critical of Russia, including accusing Vladimir Putin of ordering assassinations.

A British government investigation concluded that Litvinenko was likely poisoned by a Russian agent, probably by slipping polonium citrate or nitrate into his tea. A key was finding a radioactive tea kettle in Litvinenko's hotel room. The motive? Litvinenko's revelations about Putin's links with the underworld. It goes without saying that Litvinenko will not be getting a Gold Star as a Hero of the Russian Federation.

ELDERBERRY PLUSSES AND MINUSES

There were not many science books written back in the 1620s. Copernicus had already published his *On the Revolutions of the Heavenly Spheres* in 1543, but Galileo's *Dialogue Concerning the Two Chief World Systems* would not appear until 1632, and Newton would not lay the foundations for modern physics with his theory of gravity in the *Principia* for another six decades. However, Martin Blochwich, a young German physician, made a lasting mark with his *Anatomia Sambuci*, or in English, *The Anatomy of the Elder*, a 300-page tome dedicated to a single plant, the elderberry. The botanical name of this plant is *Sambucus nigra*, with "nigra" being Latin for black since the berries of the elderberry shrub have a deep blue-black color.

In remarkable detail, Blochwich described the botany of the plant as well as the various oils, syrups, ointments, juice, and wine that could be produced from its berries. The most appealing part of the book was a discussion of the ailments these preparations were supposed to be able to treat. Most of these treatments, such as those for tuberculosis, stomach problems, tumors, and toothache, were based on folklore rather than fact, but as it eventually turned out, elderberry extracts may have some merit when it comes to infectious diseases such as influenza and the common cold.

In 1995, Israeli researchers carried out a placebo-controlled, double-blind study that demonstrated significantly faster resolution of symptoms of influenza with a standardized elderberry extract. This was further corroborated by another such trial in Norway, this time treating sixty influenza patients with 15 ml elderberry syrup or placebo four times a day for five days. Symptoms were relieved four days earlier in the treatment group. Then in 2016, a placebo-controlled trial investigated colds in air travelers, since the frequency of infection in this group is known to be higher due to confinement in an aircraft. The placebo group participants had a significantly longer duration of cold episode days.

Exactly how elderberries deliver the goods isn't clear, but researchers have speculated that flavonoids in extracts stimulate the immune system by enhancing the production of cytokines, special proteins that cells involved in immune reactions use for communication. There is also some evidence that elderberry extracts interfere with the activity of hemagglutinin, a protein found on the surface of the influenza virus that binds it to the cell that is being infected. This suggests that elderberry preparations may be able to prevent infection if taken before exposure to the virus and can also keep the virus from spreading if taken after infection has occurred, thereby reducing the duration of symptoms.

Unfortunately, "natural health products" are poorly regulated, and in many cases, ingredients are mystical. There are numerous elderberry products available, and due to a lack of information about their composition, evaluation of potential efficacy is difficult. However, most studies that have documented positive results have used a standardized, proprietary extract that goes by the commercial name of Sambucol, originally developed by Israeli virologist Dr. Madeleine Mumcuoglu.

While commercially available standardized elderberry extracts such as Sambucol are safe, the same cannot be said for home preparations. The elderberry plant produces sambunigrin, a cyanogenic glycoside that can liberate cyanide when digested. Not enough to be lethal, but enough to cause nausea, vomiting, abdominal cramps, and weakness. Sambunigrin is found mostly in the leaves, stem, and roots of the plant but also occurs in small amounts in the seeds of the berries. If the berries are cooked, as is the case for syrups, there is no concern because any cyanide that forms will be released into the air as hydrogen cyanide. However, if elderberry juice is made by just crushing the berries without taking care to remove leaves and twigs, there can be problems. Back in 1983, eight people in California had to be hospitalized after drinking a juice made from wild elderberries and other parts of the plant that they had gathered. All recovered within a few days. Had the concoction been heated, the problems could have been avoided.

A Columbia University professor who teaches a class on contemporary civilization learned about elderberry toxicity the hard way. By her own account, she is a "great believer in natural this and that," and instead of taking a flu shot, she relied on a tincture she concocted from elderberries that she had grown to protect herself from infection by the flu virus. She became ill and ended up sending an email to her students explaining why she had to cancel class. Again, had she heated her brew, there would have been no issue. Would it have protected her from the flu? There just isn't enough evidence to say.

For protection, the flu vaccine is a far better bet, but should the flu or a cold strike, there is evidence that Sambucol can be of help in relieving symptoms. In any case, the Columbia professor hopefully learned the important lesson that "natural" does not equal "safe." Her university offers some excellent courses in chemistry. Maybe she should enroll.

TARANTULAS AND THE DANCING PLAGUE

Undoubtedly being bitten by a tarantula is not a pleasant experience. While it may cause some pain, it isn't going to cause the victim to go into a dancing frenzy. Unless the victim believes that it will! That is just what happened in episodes of "tarantism" that broke out on a number of occasions between the eleventh and eighteenth centuries in the province of Taranto in southern Italy. People who believed that they had been bitten by a type of spider, also named after "Taranto," began to dance hysterically until they were "cured" by a special type of music with an upbeat tempo, usually featuring a tambourine.

The most curious aspect of tarantism is that there is no evidence the folks affected had actually been bitten, and of course, even if they had, the result would not be a frantic dance. Even if it were, it would not be cured by music. Tarantism was actually an example of what has been called mass psychogenic illness. This occurs when large groups of people become sick because they believe that something has happened to make them sick, even though there is no actual causative agent. It is a stunning example of how the mind can rule the body. The "Tarantella" is an Italian folk dance still performed today, the origins of which trace back to the tarantism of the Middle Ages.

The Middle Ages also featured other examples of mass psychogenic illness, with the most famous one being the Dancing Plague of 1518, which broke out in what is now the French city of Strasbourg. For some unknown reason, a single woman, Frau Troffea, began to dance fervently in the street and was soon joined by about four hundred others who frolicked for days without rest until they collapsed from exhaustion. According to some reports, there were even deaths. One theory is that the bizarre behavior was caused by eating grains contaminated with ergotamine, a psychoactive compound produced by the ergot fungus. An unlikely explanation since it supposes that all those involved ate contaminated grains and that they would react the same way. It is far more likely that this was mass psychogenic illness induced by the stress of a miserable life.

More recently, in 1998, a high school teacher in the U.S. noted a gasoline-like smell in her classroom and told the students that she was experiencing a headache and was feeling nauseous, dizzy, and short of breath. Soon a number of students in the classroom began to experience similar symptoms. As the classroom was being evacuated, a number of other students and staff began to complain of similar symptoms and sought medical care. After extensive investigation of blood, air, and water samples by numerous government agencies, no toxins that could explain the symptoms were found. Authorities concluded that this was a case of mass psychogenic illness.

Another example of this phenomenon may be a condition now being called "Idiopathic Environmental Intolerance Attributed to Electromagnetic Fields." Victims claim they experience an assortment of headaches, fatigue, nausea, vertigo, difficulty concentrating, palpitations, tingling sensations, confusion, depression, and indigestion that they attribute to electromagnetic radiation from cell phones, Wi-Fi routers, or other electrical equipment. Since these people are undoubtedly suffering, there is an impetus to get to the bottom of the problem. There have now been some forty-six trials in which sufferers were asked to determine if they were being exposed to real or sham electromagnetic fields in a blinded fashion. Despite the conviction of those afflicted, these trials have been unable to replicate the claim of electromagnetic sensitivity under controlled conditions.

The researchers who analyzed all the available studies concluded that the most likely explanation for the symptoms is the nocebo effect, in which negative expectations cause symptoms even when there is no physiological reason for the symptoms. The term nocebo comes from the Latin for "I shall harm" and is the evil cousin of the placebo effect, deriving from the Latin for "I shall please." This also explains why about 20 percent of patients taking a sugar pill in controlled trials of drugs report side effects.

In yet another fascinating case, a man presented in an emergency room saying he felt terribly ill and admitted to taking a whole bottle of pills after a fight with his girlfriend. His blood pressure was very low and his heartbeat rapid. Response to treatment was poor. What sort of pills had he taken, the doctors asked? The despondent patient explained that he had been enrolled in a clinical trial for an experimental antidepressant drug and had taken all the pills he had been given. A quick call to the researcher in charge of the trial revealed that the man had been in the placebo arm of the study! When told that the pills he had swallowed contained nothing but sugar, his heart rate and blood pressure quickly returned to normal. As the poet John Milton astutely noted, "the mind can make a heaven of hell and a hell of heaven."

POWERING YOUR CELL PHONE

It is the lightest of all the solid elements, but it certainly isn't a lightweight when it comes to the punch it can pack. We are talking about lithium, an element whose compounds are at the heart of the lithium-ion batteries that power our cell phones, laptops, and electric vehicles. Little wonder that the 2019 Nobel Prize in Chemistry was awarded to professors John B. Goodenough, M. Stanley Whittingham, and Akira Yoshino for the development of such batteries.

Lithium metal reacts readily with moisture and with oxygen, so it is never found in its elemental state. For commercial production, lithium salts are extracted from water in mineral springs, brine pools, or brine deposits, found mostly in Chile and Australia, and are subjected to an electric current that converts the lithium ions to metallic lithium. This can be used to produce the lithium cobalt oxide used in lithium-ion batteries.

A typical lithium-ion battery consists of an electrode made of lithium cobalt oxide and another made of graphite, with the two separated by an electrolyte solution through which lithium ions can readily move. There is also a polyethylene barrier, just a few microns thick, that allows the passage of the lithium ions but is impervious to electrons, forcing these to travel through an external circuit.

When charging, a power supply is connected to the electrodes, sending a stream of electrons, in other words, a current, towards the graphite electrode, which then becomes negatively charged with the electrons being trapped in the graphite. At the same time, electrons are withdrawn from the lithium cobalt oxide electrode, which then releases positively charged lithium ions. These are attracted to the negative graphite electrode, where they eventually become embedded. Since graphite is an insulator, the electrons and the lithium ions in the graphite are kept from combining. Now the cell is charged.

When it comes to discharge, instead of a power source, the electrodes are connected to whatever equipment is to be run on electricity. Now the lithium ions are attracted back towards the lithium cobalt oxide electrode whence they came because being locked into a stable matrix with other metals is a more stable arrangement. The electrons, unable to flow through the barrier between the electrodes, pass through the external circuit, generating a current.

There is no question that lithium-ion batteries have changed the way the world functions. But as always, with any advancement, there is a "but." In this case, there are issues about the environmental consequences of extraction of the necessary nickel, copper, lithium, and cobalt from their ores, as well as about possible health effects when it comes to the workers involved. Then there is the problem of recycling, as well as concerns about the batteries possibly igniting and causing fires.

The extraction and processing of lithium has a large environmental footprint. It requires a great deal of water, loads of energy, and emits greenhouse gases. Recycling is also becoming a bigger and bigger issue as spent batteries from vehicles pile up. Today, only about 5 percent of batteries are recycled, with the rest ending up in landfills, potentially releasing metals that contaminate soil and groundwater. The barrier to recycling is the need for expensive, sophisticated equipment to treat harmful emissions, as well as the costs associated with the collection, storage, and transport of spent batteries.

Cobalt recycling is more critical than that of lithium since waste from cobalt mining pollutes water systems and has a number of health risks associated with it. Exposure to pulverized dust containing cobalt can cause contact dermatitis as well as respiratory sensitization, asthma, decreased pulmonary function, and shortness of breath. Most cobalt is mined in central Africa, where there is a lack of basic protective equipment such as face masks and gloves, and because of poverty, child labor is common.

Then there is fire. There are scary accounts of cell phones catching fire and airplanes crashing when lithium-ion batteries suddenly ignited. Keep in mind, though, that relative to the number of such batteries in use, the risk of fire is minute. It happens when a battery is short-circuited, usually by damage to the thin barrier that prevents electrons from flowing through the battery. If electrons do pass through, they generate a current that can ignite the pressurized, flammable electrolyte. That can occur if the battery is physically damaged, or if faulty manufacturing leaves behind tiny metal particles that are capable of causing a short circuit. Such flaws have resulted in recalls, including of some Samsung Galaxy phones.

Research into improved batteries is intense and there is no doubt that advances are on the horizon, both in efficiency and safety as well as in recycling technology. And now that I'm finished typing on my lithium-ion-battery-powered laptop, I'll make a few calls on my lithium-ion-battery-powered cell phone, including one to lithium-ion-battery-powered car manufacturers to ask about recycling.

LINCOLN'S MAGICIAN

One day in the early 1900s, Robert Todd Lincoln, son of the assassinated president, was surprised to receive a photograph from Harry Houdini, the most celebrated magician of the times. In the picture, Houdini is seen apparently conversing with Honest Abe. Since Lincoln was assassinated in 1865 and Houdini was born in 1874, the only way they could have appeared together in a photo was if Lincoln's image was a manifestation of his "spirit." Basically, the photo purported to show Houdini conversing with the ghost of the slain president!

Such spirit photos were quite the rage at the time. The Civil War had brought spiritualism to the forefront, with distraught relatives of soldiers who perished in battle hoping to communicate with their spirits. Mediums capitalized on such folly by duping the gullible with photographs that were obviously taken after a soldier's death but nevertheless clearly showed their image, which was claimed to be a spirit apparition.

Houdini was greatly perturbed by mediums who used tricks to defraud people and was a staunch foe of spiritualists. Being a great admirer of President Lincoln, he was especially angered by mediums who claimed to have made contact with the late president and attacked them with vigor. When a medium professed to have spirit photographs of President Lincoln, Houdini sent a photo of himself together with Lincoln to the president's son to ensure he would not be ensnared by the medium's chicanery. There was an accompanying explanation describing how such photos can be readily produced with double exposure techniques.

Houdini venerated President Lincoln and was an enthusiastic collector of Lincoln memorabilia. Exactly why he had become so devoted to Lincoln's history isn't clear, but it may well have been due to the president's connection to escape artist and magician Horatio Cooke during the Civil War. At the age of eighteen, Cooke joined the Union army and amazed his fellow soldiers with magic tricks and an uncanny ability to escape when tied up with ropes. Secretary of War Edwin Stanton came to hear about Cooke's prodigious feats and thought that they could be put to some use. He invited Cooke to demonstrate his talents at a meeting that, much to the soldier's surprise, included President Lincoln.

Cooke was tied up with 50 feet of rope and quickly managed to liberate himself, a feat that so amazed the president that he appointed Cooke as a federal scout with a mission of penetrating Confederate lines as a spy and sending back reports of activities. Were he to be captured, the president thought, Cooke would have a good chance of escaping.

Indeed, he *was* captured and tied to a tree to be dealt with in the morning, which likely meant execution. But Cooke managed to put his talents to use, freed himself from the ropes, and escaped. After the war he forged a career as a professional magician, calling himself Professor Harry Cooke. He also became fascinated with spiritualism and recognized that mediums were using various magic effects to convince sitters at a séance that spirits were present. Cooke was so disturbed by this that he began to bill himself as the Celebrated King of Spirit Exposers and devoted his performances to debunking spiritualists. His exposés were very effective and so enraged spiritualists that a believer shot Cooke at point-blank range. He recovered and went on to perform for another quarter of a century.

Cooke's story appealed to Houdini because he had the same disdain for spiritualists and had often been threatened by them. He went out of his way to visit Cooke, who had obtained the moniker the Oldest Living Magician in Los Angeles, and credited him in his 1924 book *A Magician Among the Spirits* for exposing the stealthy work of mediums. Cooke died that year, but at the age of eighty, in front of an assembly of magicians and wearing a Union uniform, he had reprised the escape he had performed in front of President Lincoln some sixty years earlier.

There is yet another reason for which Houdini held President Lincoln in regard. He was keenly aware of a story about the president's interest in magic! At a rehearsal for a Fourth of July parade, a magician produced an egg from the mouth of the president's son, and upon being introduced to the magician, Lincoln surprised him by saying, "Why of course, it's Signor Blitz, one of the most famous men in America." Lincoln knew that Blitz had performed at many Union Army Hospitals during the Civil War and was thankful for the magic he had brought to the soldiers.

Now you know of the magical connection between Harry Houdini and the sixteenth president of the U.S. At his stage performances, Houdini paid homage to the president with an eagle that magically appeared out of a blend of red, white, and blue streamers. The eagle's name? Abe Lincoln.

SLOWING DOWN THE CLOCK

We can't avoid aging. Every passing minute brings us one minute closer to the end. Not a pleasant thought. So, it is little wonder that the term "anti-aging" has been seized by marketers of various cosmetics, health supplements, exotic juices, and dietary regimens. "Anti-aging medicine" is a growing field, with numerous biotech companies working on drugs designed to combat the aging process.

The quest for immortality, of course, is not new. The ancient alchemists sought to turn base metals like lead into gold in order to find the secret of gold's "immortality." After all, the metal would not tarnish; it maintained its beautiful sheen and seemed to last forever. If they could find its magic, they could perhaps apply it to humans.

But they never did find the secret. Lead is still lead and the alchemists are long dead.

This is not to say that there are not some intriguing possibilities that may help slow down the clock. As we age, an increasing number of our cells enter a stage of senescence, in which they no longer divide and begin to release chemicals that cause inflammation, resulting in damage to tissues. A buildup of senescent cells, sometimes called zombie cells, is a hallmark of aging. Can anything be done to prevent this buildup? Possibly. At least in mice. When researchers at the Mayo Clinic injected just a small number of senescent cells into young mice, their speed, endurance, and strength eroded to levels seen in a senior mouse in just a few weeks. When the mice were then treated with dasatinib and quercetin, a combination of drugs known to destroy senescent cells, they recovered most of their lost physical capabilities within two weeks! Quite dramatic! But mice are not people, and while quercetin is a safe compound extracted from apple peel, dasatinib is a very expensive leukemia drug with loads of side effects. Still, this experiment is a proof of principle, demonstrating that destroying senescent cells with "senolytics" is worthy of exploration.

However, slowing aging may not be a matter of what we do, but what we don't do — eat! It may be that if we want to live longer, all we have to do is eat less. Calorie restriction has been the only surefire way that scientists have found to slow aging in animals from rodents to monkeys, and now we are beginning to accumulate data that suggests this applies to humans as well. The idea that less is more when it comes to eating is not new. Hippocrates noted that fat people were more likely to die suddenly than slender, and Avicenna, the famed Persian philosopher and physician, suggested that the elderly should eat less than when they were younger.

Venetian nobleman Luigi Cornaro may have been the first to put a restricted-calorie diet to a test in the seventeenth century when he came to believe that his health was deteriorating due to excessive partaking of food, drink, and sex. He then restricted himself to no more than 350 grams of food a day and 400 milliliters of wine and lived to the ripe old age of ninety-eight. He documented his regimen in his book *Discourses on a Sober and Temperate Life* and described how the changes he made in his lifestyle allowed him to remain in vigorous health well into old age.

Today, members of the Calorie Restriction Society, which has the goal of increasing longevity, are following in Cornaro's footsteps and are acting as human guinea pigs. They consume no more than 2,000 calories per day, which is just over half of what an average North American wolfs down. They don't do this though by eating half servings of hamburger, fries, or pizza. They do it by following a diet high in whole grains, fruits, vegetables, beans, and fish. And according to a study published in the *Journal of the American College of Cardiology*, the austere regimen is paying off. Researchers examined the heart function of twenty-five members of the Calorie Restriction Society and to their amazement found that the hearts functioned like those of people fifteen years younger.

But permanent hunger is not appealing. Could there be some way to mimic the effects of calorie restriction without going hungry? That possibility emerged when Harvard geneticist Dr. David Sinclair showed that resveratrol, a compound found in red wine, was capable of activating a gene that codes for the production of proteins known as the sirtuins, which trigger a survival mechanism that can extend life. Furthermore, it was determined that calorie restriction in animals also results in the activation of the gene that codes for sirtuins. The media was quick to jump, and articles appeared about the purported benefits of red wine being justified by the discovery of resveratrol activating this "longevity gene." Not mentioned was that the dose of resveratrol showing benefits in the animal studies was equivalent to that found in roughly a hundred bottles of wine.

However, since resveratrol is not difficult to synthesize, the prospect of developing a pill containing a sufficient amount of resveratrol, or perhaps an even more active derivative, was raised. Could a pill be the long-sought-after fountain of youth? Despite over 20,000 research papers published on various aspects of resveratrol, no human trials have made a convincing case for resveratrol supplements, although safety, even with large doses, has been demonstrated. There is no magic bullet to slow down aging. Paying attention to diet, a good program of exercise, and early detection of disease is the best we can do. And of course, proper selection of one's parents.

KING TUT'S TOMB

February 17, 1923, marked one of the most important dates in the history of archeology. In the presence of twenty witnesses, archeologist Howard Carter unsealed an intact ancient Egyptian tomb he had discovered in the Valley of the Kings three months earlier. Eyes feasted on a throne covered in gold leaf, but it was a stone sarcophagus and its beautifully gilded inner coffins that would get the most attention. Inside was the mummy of "the boy king" Tutankhamen, adorned with the magnificent gold mask that has since become famous around the world.

There were over 5,000 other items found in the tomb, including food, wine, archery bows, and fresh linen underwear. It would not be fitting for the spirit of royalty to appear in the Hall of Judgment hungry, thirsty, and sporting dirty garments. It was in that hall that Osiris, the god of afterlife, lord of the dead and rebirth, assessed the purity of a person's heart. The Egyptians believed that the heart, not the brain, was the essence of a person and its "purity" reflected lifetime deeds. If the heart was judged to be pure, the spirit was allowed to enter immortal paradise. If not, the heart was forfeited to Ammit, the goddess who was the Devourer of the Dead, who was ready to swallow the tainted heart, thereby condemning the soul to a second death, from which there was no escape.

The Egyptians did not think that the body would literally come back to life; after all, there was plenty of evidence that this did not happen. The curse of mummies wandering around with outstretched arms, ready to torment the living, was a creation of twentieth-century filmmakers. What the Egyptians did believe was that the spirit had to regularly return to the body for sustenance, hence the importance of preserving the corpse through mummification. Funeral masks were placed over the deceased's face so that the spirit could recognize the body upon its return. Since the home of the spirit was the heart, it was preserved and replaced in the body cavity. The unimportant brain was removed with special tools through the nose and discarded.

A variety of amulets and good-luck charms, which were blessed by priests with special incantations to offer protection against evil and unhappiness, were often included with the mummy. The most widely used amulets were replicas of scarab beetles, now commonly referred to as dung beetles. To the ancient Egyptians, these insects were sacred because of their practice of collecting dung and rolling it into a ball. This seemed akin to the ritual followed by Khepri, the god who was believed to roll the sun across the sky every day, just like a dung beetle rolls dung. Indeed, Khepri was often represented as a scarab-headed man. The scarab was also seen as a symbol of rebirth and regeneration since it lays its eggs in the ball of dung from which its offspring eventually emerge.

Scarab replicas were carved out of stone or ivory, or formulated from an early form of ceramic made by mixing sand or crushed quartz with sodium carbonate (soda) and calcium oxide (lime). When fired at a high temperature, this mixture vitrifies and develops a glaze on the surface that can take on a blue or green color if copper compounds are incorporated into the mix. Scarabs were worn as necklaces, distributed around the home, and were also placed on the chest of a deceased person just over the heart to prevent any sins that had been accumulated in life from being released when Osiris weighed the heart for purity. Tutankhamen had a large scarab carved out of stone on his chest.

Just how the purity of Tutankhamen's heart may have been judged we will never know. Aside from some minor restorations of temples, he does not seem to have accomplished much in his young life, but then again, he was only on the throne for nine of his seventeen years. On the other hand, there is no evidence that he would have done anything to defile the purity of his heart. While he didn't do much in life, in death Tutankhamen has become probably the most famous of the pharaohs. This is due partially to the golden mask that has been exhibited and admired around the world as well as the supposed curse that was inflicted on all who tampered with his tomb.

That curse, basically the invention of newspapers seeking publicity, supposedly resulted in the premature death of people who were in any way associated with the opening of the tomb. The fact is that of the fifty-eight people who were present when the tomb and sarcophagus were opened, only eight died within a dozen years; Howard Carter died of lymphoma in 1939 at the age of sixty-four. There was nothing unusual about the deaths. There was no pharaoh's curse.

UP IN SMOKE

The kits posted along the banks of London's Thames River in the late eighteenth century contained emergency equipment for the resuscitation of drowning victims. Along with smelling salts, there were wooden pipes to blow air through the nostrils into the lungs. But the kit also contained a flexible rubber tube, a set of bellows, and a combustion chamber for burning tobacco. If blowing air into the lungs failed to produce results, then the next step was to use the bellows and tube to introduce tobacco smoke. Not through the mouth or nose, but through the rectum! A delicate procedure to be sure.

In 1774, physician Thomas Cogan and pharmacist William Hawes founded the Institution for Affording Immediate Relief to Persons

Apparently Dead from Drowning and fostered the notion that the administration of tobacco smoke rectally stimulates the heart, encourages respiratory function, and dries out the waterlogged insides. Needless to say, no drowning victim was ever resuscitated in this fashion.

Cogan and Hawes were not the first to claim benefits for blowing tobacco smoke up the rear. In fact, when Columbus arrived in North America he found that natives were using tobacco for the treatment of various ills. Whether they pioneered the tobacco-smoke enema, as some accounts claim, isn't clear. However, it is a fact that after tobacco was introduced into England, smoke "per rectum" was prescribed as a treatment for conditions as varied as constipation, headaches, strychnine poisoning, worms, hemorrhoids, typhoid fever, and cholera. Before bellows were introduced, physicians would use their mouth to blow tobacco smoke up the rectum through a tube. This could have disastrous results if the practitioner inadvertently inhaled while administering treatment.

The greatest danger from tobacco, however, was not from futile medical treatments, but from smoking for pleasure. Somewhat surprisingly, this risk did not hit the headlines in a major way until British epidemiologist Richard Doll published his now-famous study in 1950. Doll, a physician, had developed an interest in epidemiology, the branch of medicine that studies the occurrence and possible causes of disease in different groups of people.

During the first half of the twentieth century, doctors had noted an increase in the rate of lung cancer, and Doll with colleague Austin Bradford Hill undertook a study to investigate why this was the case. They questioned 700 lung cancer patients in twenty London hospitals, and before long discovered that the most prevalent common feature was that they were smokers. Halfway through the questioning, Doll became so convinced of the link between lung cancer and smoking that he gave up the habit!

The study was published in 1950 in the *British Medical Journal*, noting that the risk of developing lung cancer increases in proportion to the amount smoked, and may be fifty times greater among those who smoke twenty-five or more cigarettes a day than among non-smokers. In 1951, the two researchers went on to organize a prospective study by writing to all registered physicians in the U.K. asking if they would be willing to periodically fill out questionnaires about their lifestyle and health status. Some 40,000 responded, and by 1954 it had become clear that smoking and lung cancer were linked.

The results of Doll's study prompted Health Minister Iain Macleod to announce at a news conference in 1954 that "it must be regarded as established there is a relationship between smoking and cancer of the lung." Curiously, Macleod didn't seem to be too bothered by the relationship given that he chain-smoked through the whole press conference!

Responses from the physicians were collected intermittently until 2001 with even more revelations. Heart attacks occurred more frequently in smokers, and smoking seemed to decrease life expectancy by as much as ten years, with 50 percent of all smokers dying of a smoking-related disease. In addition to lung cancer, kidney, larynx, neck, breast bladder, esophageal, pancreatic, and stomach cancers were also implicated. Emphysema, strokes, chronic bronchitis, premature births, and high blood pressure are also more likely in smokers.

Given the current state of knowledge, it comes as no surprise that many smokers would like to kick the habit. Electronic cigarettes were originally introduced as a safer alternative since they do not produce the various carcinogens and irritants found in tobacco smoke but still satisfy cravings for nicotine. However, it is now clear that "vaping" exposes users to a host of compounds originating from the decomposition of the propylene glycol, glycerol, vitamin E acetate, THC, nicotine, and various flavor components e-cigarettes can contain. As evidenced by recent cases of respiratory ailments, vaping is not benign. While it may help some to give up smoking, there is concern that it may also lead young people to experiment with cigarettes. Then there is the further issue that the readily available refills contain a potentially lethal amount of nicotine if ingested.

Warnings about vaping should not be considered as just blowing smoke up the . . .

KAMBO

It is not a pretty scene. In the heart of the Amazon a giant leaf frog, also known as a giant monkey frog, is spread-eagled in an undignified X shape, with each of its limbs secured to a wooden stick stuck into the ground. A shaman then proceeds to scrape a waxy secretion from the amphibian's back and applies a drop to a small wound inflicted with a burning stick on a person's arm. Within minutes the recipient experiences projectile vomiting. Just picture the classic scene in *The Exorcist*. Sometimes other body orifices also get in on the action. After the torturous purging has subsided, subjects say they feel peaceful and refreshed. And, undoubtedly, relieved.

This traditional ceremony is called kambo, and the same term is used to describe the mix of chemicals that the frog with the scientific name *Phyllomedusa bicolor* secretes to deter predators. Kambo has a long folkloric history among Indigenous tribes of the Amazon; it is said to boost strength, increase stamina and alertness, bring good luck to hunters, cure disease, and cleanse the body and soul.

Recently the kambo ritual has emerged from the rainforests of the Amazon to take on a role in Western society as a form of "alternative" medicine. As disenchantment grows with the failure of conventional practice to cure all diseases, people are being increasingly attracted to "natural" substances produced by plants or animals. Isn't it curious that people who worry about taking a "synthetic" pharmaceutical backed by decades of research will uncritically subject themselves to the poisonous secretions of an Amazon amphibian?

Yet that is just what is happening at ceremonies known as kambo circles. These are not conducted by expatriate Amazon shamans but rather by self-taught individuals who may or may not be members of the International Association of Kambo Practitioners, which offers accreditation in the practice. It isn't clear what accreditation involves. It certainly doesn't involve applying evidence-based science, because there isn't any to support this bizarre, albeit legal, ritual.

Why then do people allow some practitioner with no medical training to introduce a mix of toxins into their body through freshly burned holes in their skin? It is not because of a quest for euphoria, nor a desire for hallucinogenic effects, because kambo produces neither. Some individuals, driven by desperation, are in search of a solution for their pain, depression, or addiction based on some anecdotal accounts they have come across. Most, however, come seeking to "purify their body" by cleansing it of unidentified toxins.

The purge that begins within minutes of administering kambo is often accompanied by an increased heart rate and sweating, described by one victim as "like having the flu and food poisoning multiplied a hundredfold." Why is it tolerated? Because the expulsion of vomit is seen as ridding the body of spiritual and emotional baggage!

After the symptoms subside, people claim to feel increasingly energetic and in possession of greater "mental clarity," whatever that means. There are also claims of boosting the immune system, relieving allergies, curing headaches, and alleviating sadness, none with any corroborating evidence.

It should come as no surprise that kambo has physiological effects given that it is a toxic secretion meant to make predators sick enough to swear off trying to make a meal of the large green frog ever again. While not all the components of kambo have been identified, chemists have managed to isolate a number of bioactive peptides. These are short chains of amino acids that can have various effects including stimulating motility in the gastrointestinal tract, boosting adrenaline production, and affecting blood pressure and heart rate. Some, referred to as opioid peptides, have an affinity for opiate receptors and can, in theory, alleviate pain, but the concentrations in kambo are too small to have any biological activity in humans.

As far as safety goes, there have been a couple of case reports of death following the use of frog toxins, but whether these actually played a role isn't clear. There is, however, one documented case of a woman experiencing lethargy, confusion, cramps, and finally a seizure after a kambo experience. She had consumed six liters of water, as she was told to do to "flush out the toxins," but what she actually ended up flushing from her body was sodium, resulting in a case of hyponatremia. This was due to too much water intake coupled with a reduction in the production of a hormone that prevents excessive urination, an effect linked to one of the kambo peptides. Treatment with salt and restriction of water intake solved the problem. Still, relative to the number of people who have engaged in frogging rituals, the risk appears to be small. Even for the frog. The creature is released, apparently none the worse for wear.

Although exploring the chemistry of frog venom is intriguing, I'm not partial to projectile vomit. I think I would rather boost my spirits with another product of the Amazon. *Theobroma cacao*.

MUSTARD GAS AND CHEMOTHERAPY

It makes for a compelling story. The Second World War is being furiously fought across Europe. The Allies finally gain a foothold in Italy and the port city of Bari becomes a critical point of entry for troops and supplies to the Mediterranean theater. The harbor is filled with ships on December 2, 1943, when Nazi airplanes drop from the clouds, their bombs raining destruction.

The SS *John Harvey*, an American ship carrying a cargo of 2,000 mustard gas bombs in spite of the Geneva Protocol of 1925 that banned the use of chemical weapons, explodes, killing all of its crew and spreading the gas across the harbor and town. But the clouds from which the Nazi planes emerged have a silver lining. Researchers note that the victims of mustard gas exposure have a very low rate of white blood cell multiplication, suggesting that mustard gas could also interfere with the characteristically rapid multiplication of cancer cells. And so it is that the Bari attack serendipitously leads to the development of mustard gas as an anticancer drug and launches the concept of chemotherapy. At least that is the way the story is told in numerous textbooks and articles.

A nice romanticized account, but the fact is that the first use of a modified version of mustard gas to treat cancer in a human was in the U.S. in 1942, more than a year before the Bari attack! The seminal event that gave rise to the treatment was indeed a mustard gas attack, but one that the Germans unleashed on Allied troops at Ypres in Belgium in 1917.

Mustard gas, which is actually a yellowish liquid that evaporates readily, was first made in 1822 at the École Polytechnique in Paris by Professor César-Mansuète Despretz, although he did not comment on any irritating properties. Those were noted by German chemist Albert Niemann in 1860 and later documented in rabbits by Victor Meyer, who along with Hans T. Clarke developed synthetic methods appropriate for large-scale production. While working in famed professor Emil Fischer's lab in Berlin, Clarke broke a flask of mustard gas and suffered burns to his leg that required two months of hospitalization. It is believed that it was Fischer's reporting of this incident to the German Chemical Society that planted the seed for Germany's use of mustard gas in the First World War. Another factor that played a role was the availability of a supply of ethylene chlorohydrin required for mustard gas production. This was thanks to the large dye manufacturing industry that had been established in Germany, which required the use of the chemical.

Mustard gas readily penetrates skin, causing blisters and burns. If inhaled, damage to the respiratory system can be fatal. Survivors of a mustard gas attack are not out of the woods because the gas damages DNA, resulting in aplastic anemia, the decreased formation of blood cells by the bone marrow. This was first documented in "Blood and Bone

Marrow in Mustard Gas Poisoning," a journal article by American physician Dr. Edward Krumbhaar, who had served in France as a medical officer and conducted autopsies on victims of the 1917 Ypres attack.

Prior to the Second World War, Yale pharmacologists Alfred Gilman and Louis Goodman had been recruited by the U.S. War Department to develop an antidote for mustard gas. Realizing that this was a dangerous chemical, they looked for less-problematic analogues for their experiments. Substituting a nitrogen atom for sulfur in the molecular structure resulted in mechlorethamine, a "nitrogen mustard" that was easier to handle. Being familiar with the work of Dr. Krumbhaar, they now wondered if this novel compound would have an effect on cancer cell proliferation similar to the one the original mustard gas had on white blood cell multiplication. They soon showed that indeed it did! A mouse that had received a transplanted lymphoma tumor was the beneficiary. When treated with the nitrogen mustard, the tumor softened and regressed.

Now it was time to test the treatment on a human. A forty-eight-year-old bachelor, only identified by his initials, "J.D.", suffered from huge tumors that affected all his movements as a result of non-Hodgkin's lymphoma. Radiation and surgery had short-term effects, and he was asked if he would be interested in volunteering for an experimental treatment. J.D. jumped at the chance.

Daily injections of mechlorethamine quickly reversed the symptoms, and within a month the tumors had vanished. Unfortunately, the cancer soon returned and subsequent treatments failed to have an effect. Three months after starting treatment, J.D. died. But the possibility of treating cancer with drugs had been demonstrated and J.D.'s case was a springboard for developing a range of chemotherapeutic drugs. This was more than a year before the Bari attack!

This is not to say that the Bari incident played no role in the history of chemotherapy. After the Bari attack, hundreds of military and civilian personnel were hospitalized, many of whom ended up dying. An American army doctor, Lieutenant-Colonel Stewart Francis

Alexander, was put in charge of investigating the health consequences of exposure to mustard gas and confirmed its effect on reducing cell multiplication. This added impetus to the research, but the Bari event was not the key to the development of chemotherapy, albeit that is how it is often portrayed. That dubious "credit" goes to the 1917 mustard gas attack at Ypres.

WANT A DATE?

It's tough to be a lab rat. Learning to navigate a maze or having your anogenital distance measured to monitor exposure to endocrine disruptors is unpleasant enough, but not nearly as troubling as having one of your testicles twisted 720 degrees so that researchers can study the effects of antioxidants on testicular torsion. In men, testicular torsion is a medical emergency that requires quick treatment. This occurs when the spermatic cord, which provides blood flow to the testicle, rotates and becomes twisted. As a result, the testicle's blood supply is cut off, causing sudden pain and swelling. If a physician cannot resolve the problem by manual manipulation, the patient is off to surgery. Untwisting the spermatic cord restores the blood flow but introduces another problem. The sudden influx of oxygenated blood leads to the formation of free radicals that can lead to tissue damage. The presence of antioxidants can conceivably prevent such damage.

That is exactly what researchers were interested in investigating. They chose an extract of dates as a source of readily available antioxidants, but they could have chosen a variety of fruits or vegetables. Rats were treated with the date extract through a tube introduced into their stomach and subsequently had one of their testicles surgically rotated. After torsion had been introduced, the problem was resolved surgically and the effects on the testicle were studied and compared with that in rats that had not been pretreated with the date extract. The researchers report that the date extract had a protective effect.

What does one take away from such a study? Certainly not that men should be walking around with dates in their pocket to be consumed in case testicular torsion presents. Neither can one conclude that there is something special about dates; all fruits and vegetables contain an array of antioxidants. The only message is that it may be worthwhile to study whether treating men who are about to undergo testicular detorsion with antioxidants yields a benefit. Of course, it would be easy to fabricate a sensational headline from this study. For example, "Dates can prevent testicular damage in men" would certainly garner attention. While I haven't actually seen such a headline, I have seen articles that talk about curing cancer with dates.

It seems these are based upon a paper published in the *International Journal of Clinical and Experimental Medicine* with the title "Therapeutic effects of date fruits in the prevention of diseases via modulation of anti-inflammatory, antioxidant and anti-tumor activity." The notion that the sweetest fruit in existence would produce such sweet science is seductive, but I would suggest that this paper does not provide evidence to justify this title. The authors begin with a statement that the current treatment of cancer and diabetes is expensive and has adverse effects, and that an alternative approach that is safe, effective, and affordable is needed. Nobody would contest that. Then, they go on to say that natural products are a good remedy as they are inexpensive and easy to access without complications. That may be, but what about efficacy?

The researchers, who hail from Qassim University in Saudi Arabia and Suez Canal University in Egypt, attempt to make a case for the therapeutic effects of dates by surveying the literature for studies that investigate any of the natural components of dates. Like any fruit, dates harbor literally hundreds of compounds, including sugars, fats, proteins, vitamins, minerals, polyphenols, carotenoids, glucans, steroids, and saponins. Study any specific component of dates, or some extract of the fruit in the lab or in an animal model, and some effect will be noted. You can then torture the data until it reveals some potentially therapeutic effect. There are hundreds and hundreds of compounds that have been isolated from botanicals that have shown antioxidant, anti-inflammatory, and anticancer effects in the lab. Only in extremely rare instances have these made the jump to human therapeutics. For every Taxol or vincristine, there are thousands of plant compounds that have ended up in pharmacology's junk pile.

The fact that some date components or date extracts may neutralize free radicals in a test tube or interfere with the multiplication of cancer cells says nothing about how consuming dates may affect any disease. Neither does injecting purified beta glucan extracted from dates into mice that have had tumors transplanted into them and noting tumor regression tell anything about the value of eating dates.

So, while dates may not delay a date with the Grim Reaper, they do have something going for them. Taste! Especially the Medjool variety, regarded as the "king of dates," thanks to their large size, soft texture, and rich, sweet flavor. These dates are packaged as soon as they are harvested and, unlike most other dates, are not dried. Don't go overboard though. Dates have the highest sugar content of any fruit, with four dates containing about 60 grams, close to double that in a 12-ounce soft drink!

FLUORIDE AND IQ

A classic scene in the 1964 film *Dr. Strangelove* has General Jack D. Ripper claim that fluoridation is a "monstrously conceived and dangerous communist plot." While obviously a spoof, the movie nevertheless made fluoridation a topic of discussion. That discussion has continued unabated, with proponents claiming that the fluoridation of water is one of the most important public health achievements, and opponents arguing that putting "rat poison" into water undermines health.

It comes as no surprise that a study by researchers at York University captured the media's attention after it raised the prospect of a connection between fluoride intake during pregnancy and lower IQ among offspring. Published in the highly respected *Journal of the American Medical Society Pediatrics*, the paper received unusual editorial scrutiny because of concern about the alarm such a study could raise. The editor took the extraordinary step of consulting more reviewers than usual, all of whom agreed that data collection was sound and the statistical analysis correctly carried out. Of course, as usual with such epidemiological studies, there are caveats.

Previously scientists in Mexico City had found a significant correlation between reduced IQ in children and urinary fluoride during pregnancy, and a study in China revealed lower IQ in children who were consuming water high in naturally occurring fluoride. The current study aimed to compare the IQs of children born to mothers who live in regions of Canada where the water supply is fluoridated with those born to mothers living in non-fluoridated areas. The first finding that jumps out is that the IQs of the children in the two groups are identical (108), so there is a question of why further investigation was even undertaken. It was only with "data mining" that some differences surfaced.

There were two arms to the study. One looked at urinary fluoride output, and the other at ingested fluoride by pregnant women. Urinary fluoride was found to correlate with lower IQ in the offspring at three to four years of age, but only in boys, which is very curious. When it comes to ingested fluoride, there was an association with lower IQ in both boys and girls, although it was barely statistically significant. The problem, though, with ingested fluoride is that it is hard to assess, especially when the evaluation is through recall via questionnaires. It is a tough task to remember how much tap water one consumed, and virtually impossible to know whether foods and beverages consumed away from home contain fluoride. Basically, what we have is a guesstimate. A further problem is that urinary fluoride did not correlate well with ingested fluoride. That is a conundrum and calls into question the validity of the ingested fluoride estimate.

Based on the data, the researchers conclude that a 1 milligram increase in daily maternal fluoride intake is associated with a deficit of

3.7 points in IQ. While this extrapolation is mathematically correct, it is misleading because the mean fluoride intake in the group that lived in non-fluorinated areas was 0.30 milligrams per day and 0.93 in the group that drank fluoridated water. So, there is no 1 milligram increase! The difference is 0.63 milligrams, which would lead to a much smaller reduction in IQ than 3 points, a difference that is likely of no practical significance.

The researchers did control for possible confounding factors, but there was no control for the mothers' IQs. Other possible influencers such as exposure to lead, or fluoride intake by the children between the time they were born and the time of IQ assessment, were not taken into account. Also, the half-life of fluoride in the body is short, so urine levels depend on when the sample is taken relative to the time that fluoride-containing foods or beverages were last consumed.

In science, while a single study can add to the body of knowledge about a subject, it does not warrant making wholesale lifestyle changes until it is reputably repeated. In any case, based on the data unearthed here, one can conclude that if there is an effect on the IQ of children as a consequence of fluoride intake during pregnancy, it is small, and of questionable significance. However, any prospective mom who is concerned can easily replace fluoridated tap water with bottled water for a few months. This is very unlikely to make a difference in dental health, especially if fluoridated toothpaste is used.

When it comes to risk factors that can affect a fetus, I would rank tobacco smoke, alcohol consumption, air pollution, and exposure to lead, mercury, cadmium, and various potentially endocrine-disrupting chemicals such as phthalates well ahead of worries about fluoride.

While this recent study about IQ and fluoride is noteworthy and should trigger further investigation, it leaves us with only one certainty. Fluoridation supporters and anti-fluoride activists will spin the data to claim either that this study proves nothing or that water fluoridation should be discontinued because it has a dumbing-down effect on the population.

THE SCIENCE AND PSEUDOSCIENCE OF ESSENTIAL OILS

I try to avoid walking through the ground floor of department stores. I find the cacophony of smells emanating from the various perfume counters literally tear-jerking. My eyes water until I manage to make my escape. It's an annoying business, but it does make me reflect on the chemistry of the essential oils that are the source of many of the compounds floating about. In this case, the term "essential" means that the oil contains the essence of a plant's fragrance, not that it is in any way essential for its existence or for the existence of any other living organism. Essential oils are used to produce perfumes and to add scent to cosmetics and cleaning products. They are also used as flavorings in foods and beverages and have been used historically as medical treatments by application to the skin, through ingestion or through inhalation. The latter is commonly referred to as aromatherapy.

In some cases, as for citrus fruits, the essential oil can be isolated just by mechanical expression — in other words, squeezing of the peel. Solvent extraction is another path to essential oils. Treating plant material with alcohol, hexane, or liquid carbon dioxide extracts a mixture of organic compounds that are left behind as the essential oil when the solvent is removed by evaporation. However, the most efficient and widely used method to isolate an essential oil is distillation.

Distillation was introduced sometime during the twelfth century and involves heating a mixture of substances and condensing their vapors into a liquid. Compounds with lower boiling points can be separated from ones that do not boil readily. One problem is that many plant compounds decompose at high temperatures. Steam distillation, a technique that introduces water or steam into the distillation apparatus, is a way around this problem given that a mixture of water and organic compounds boils at a lower temperature than either component individually. The essential oil can be easily separated from the water that is also distilled since water and oil do not mix.

Now we move on from the science of essential oils to the pseudoscience. And there is plenty of it. Depending on which pseudo-expert has mounted the soapbox, either sniffing the right essential oil or rubbing it on the skin can support the immune system, enhance mood, promote sleep, cleanse the body's organs, boost the libido, ease breathing, foster alertness, treat kidney stones, oxygenate the blood, relieve pain, reduce anger, prevent constipation, and, of course, eliminate toxins. If that isn't enough, essential oils are also reputed to readjust chakras, harmonize bioelectrical frequencies, cleanse negative energies, drive out evil spirits, and promote sexual stimulation. In case you are interested, the latter involves massaging the appropriate area with jasmine oil. That is likely to work whether you use an oil or not.

Needless to say, according to enthusiasts, essential oils must be composed only of naturally occurring compounds; synthetics need not apply for inclusion. Why? Because natural substances possess some sort of "life force" absent in synthetics, a claim that was buried two centuries ago with Friedrich Wöhler's demonstration that the urea he had synthesized in the lab was identical to the natural version isolated from urine.

Sales of essential oils are dominated by multilevel marketing (MLM) companies that snare potential participants with promises of wealth through a commission system. Unfortunately, this often drives individuals to make outlandish claims about using the oils to treat cancer, autism, Alzheimer's disease, mononucleosis, or arthritis. There seems to be an oil for any condition that potential customers have. The Food and Drug Administration in the U.S. has sent warning letters to the major MLM companies, resulting in more careful wording of claims, but there is no way to police what parties say in the privacy of a home, where most sales are made.

Of course, just because some claims on their behalf stink, does not mean that essential oils are useless. The scent of lavender seems to have a calming effect on some people and helps with sleep, but it can cause headaches in others. Peppermint oil may be of some use in indigestion, but that is through ingestion, not inhalation.

Still, further research into aromatherapy may lead to some interesting applications. Dr. Alan Hirsch, a legitimate expert in olfaction, measured penile blood flow in thirty-one male volunteers who were either wearing scented masks or non-odorized masks. The greatest increase in blood flow was seen with the combined odor of lavender and pumpkin pie. And wouldn't you know it? A company has already jumped on that bandwagon and is marketing Pumpkin Lavender Perfume Oil. May be worth an experiment.

There's no question that a massage with an essential oil or soaking in a scented bath may have a pleasant, relaxing effect, but it is not going to "align your DNA," "repair your energy field" or "keep your nerves in balance." Actually, the only thing such claims can do to nerves is fray them.

BUG JUICE

Any youngster who has been to summer camp will be familiar with "bug juice." That's the drink made by dumping fruit-flavored crystals and sugar into water. Why bug juice? There are a couple of theories. At camp, the concoction is often made in large buckets, and the sugar attracts flies and other bugs. Hence "bug juice." But there is another contender for the expression.

Carmine is a popular red food dye that is extracted from the female cochineal insect. Over the years there have been all sorts of fruit-flavored drink mixes that were colored with cochineal extract, conceivably giving rise to the bug juice moniker. Amusingly, there is actually a commercial drink called Bug Juice; however, it is colored with red dye #40, not cochineal.

While the origin of the term may be up in the air, it is safe to conclude that nutritionally bug juice doesn't have much going for it.

In the future, though, we may be serving up another type of bug juice, one made from insects, one that may help feed the 9 billion people who will be coming to dinner by 2050. Keeping massive hunger at bay will require a significant increase in food production, with some researchers looking to entomophagy as an attractive possibility.

The term derives from the Greek *entemon* for insect, and "phagein," "to eat." Insects can be very nutritious, are relatively easy to raise, and have a much smaller environmental impact than animal agriculture. A kilogram of crickets provides about 200 grams of protein, not much less than the 250 grams found in a kilogram of beef. Furthermore, cattle require about 8 kilograms of feed to produce 1 kilogram of meat, while crickets can produce the same amount from just 2 kilograms of feed. Raising crickets instead of cattle requires less land for growing feed since insects are not particular about their diet and will eat any sort of plant, fruit, or vegetable waste that cattle would not eat. Insects also reproduce quickly and only release a tiny fraction of the greenhouse gases produced by cattle. The downside is that in general, our mouths do not water at the thought of eating insects.

However, a study of the antioxidant potential of various creepy crawlies by researchers at the University of Teramo in Italy may add some appeal to dining on the little pests. Antioxidants have great commercial appeal, although the surrounding hype generally outdistances the science. The rationale for the importance of these chemicals is that they are capable of donating electrons to electron-poor species, such as the notorious free radicals produced as byproducts of normal metabolism.

Generally, these rogue species are kept in control by naturally occurring antioxidants such as vitamin C, vitamin E, and glutathione, all capable of satisfying the free radicals' hunger for electrons. Should that hunger not be satisfied by antioxidants, the free radicals will then endeavor to steal electrons from other molecules such as proteins, fats, or nucleic acids. Since electrons are the "glue" that hold molecules together, free radicals can tear these important biomolecules apart, resulting in a variety of health issues. That process is termed "oxidative stress" since oxidation is defined as a loss of electrons, which is just what is happening to these biomolecules. In other words, they are being "oxidized." Free radicals therefore can be referred to as "oxidizing agents" and any species that neutralizes them as "antioxidants." Oxidative stress occurs when there are more free radicals being produced than can be mopped up by antioxidants.

The Italian scientists studied the antioxidant potential of extracts of a variety of insects such as ants, grasshoppers, caterpillars, silkworms, and crickets. They actually didn't stop at insects; they even looked at tarantulas and scorpions. Surprisingly, the antioxidant capacity of crickets, caterpillars, silkworms, and grasshoppers was greater than that of orange juice or olive oil, both of which are good sources of antioxidants.

That finding was seized upon by the media, with articles suggesting that incorporating insects into the diet could protect against heart disease and cancer since free radicals have been implicated as a causative factor in these conditions. Sensation-seeking headlines such as the *Telegraph*'s "Eating Ants Could Protect Against Cancer" were clearly unreasonable.

First, there is precious little evidence that an intake of antioxidants offers protection against cancer. While there is evidence that plant-based diets are associated with a reduced risk of cancer when compared with meat-based diets, it is not clear that it is antioxidants that are responsible. Plants contain hundreds of potentially beneficial compounds, and the answer may actually lie not in what vegetarians and vegans are eating, but in what they are not eating, namely meat.

Second, there is the question of amounts. While a hundred grams of orange juice may have the same antioxidant capacity as a hundred grams of ants or grasshoppers, it is far easier to consume a hundred grams of juice than a hundred grams of insects. I know. I've tried it. Maybe putting the bugs in a blender makes for easier consumption, but so far, I have not been able to enlist volunteers to try this antioxidant "bug juice."

THE CRAZY STORY OF LOCOWEED

At first the horses in the pasture were just stumbling about in circles. Then came excessive drooling and lethargy, with occasional fits of aggressive behavior. The farmer recognized the symptoms. Locoweed poisoning!

There are some 300 species of this hardy perennial plant, a number of which produce swainsonine, the toxic alkaloid that is responsible for making the animals act "loco." That's Spanish for "crazy." Cattle, horses, sheep, goats, and wildlife are attracted to locoweed in the early spring and late fall because the weed is green while other plants are brown. Estimates are that every year livestock producers lose over $300 million due to the death of poisoned livestock. There is no treatment for "locoism," and farmers who allow their animals to graze in pastures have to scout for locoweeds and try to control them with herbicides. It is difficult to eliminate locoweed because the seeds of the plant can lie dormant in the soil for years.

Plants require a number of nutrients for proper growth, selenium being one. Locoweeds are unusual in that they require more selenium than other plants and indeed an abundance of locoweed indicates high selenium levels in the soil. In large doses selenium is toxic and can present a problem independent of swainsonine for grazing livestock.

People, like plants, also require selenium for health. Not a lot, only about 50 micrograms a day. The selenium is incorporated into some two dozen selenoproteins that play critical roles in reproduction, thyroid hormone metabolism, and DNA synthesis, as well as protection from oxidative damage and infection. Getting enough selenium is generally not a problem since it can be found in muscle meats, seafood, grains, nuts, and dairy products. However, deficiency can occur in areas where the soil is lacking in selenium.

Starting around 1935, scientists noted an unusual number of deaths due to a weakening of the heart muscle in the Keshan region of China. By 1967, Keshan disease, as the condition came to be called, was found to be more widespread but still limited to certain geographic areas. People who had once lived in these regions but then moved away were unaffected. This initiated sampling of the soil, food, and water in the areas where the disease was prevalent, as well as comparison of hair samples from people who lived in the disease-prone areas with those who lived elsewhere.

It turned out that concentrations of selenium in the soil, in the food, and in people's hair were lower in the areas affected by Keshan disease. This raised the possibility that selenium supplements could be instrumental in offering protection. After animal studies had shown that sodium selenite could be safely used, residents were given the supplement under medical supervision. Within ten years, enough data had been collected to show that such supplementation was effective for reducing the rate of Keshan disease.

In North America, diet surveys have shown that people get enough selenium to prevent overt deficiency. However, some studies have suggested that people in the highest category of selenium intake have a lower risk of some cancers than those in the lowest category. Randomized controlled trials of cancer prevention using selenium supplements have had mixed results, with some suggesting a benefit, others not. No risks associated with supplements in the range of 200 micrograms a day have been noted. One possibility of overdosing on selenium is from overconsumption of Brazil nuts, which can contain as much as 90 micrograms per nut. Toxic effects start at about 700 micrograms, so while a couple of Brazil nuts a day is fine, more than eight could create a problem. Early indicators of excessive selenium intake are a garlic breath, metallic taste in the mouth, nail brittleness, and hair loss.

Selenium had important applications long before its effects on health were noted. The element was discovered by the famed Swedish chemist Jacob Berzelius in 1817 as a contaminant in the iron sulfide he was using in the production of sulfuric acid. Because of its similar properties to tellurium, an element discovered some twenty years earlier, and named after the Latin for Earth, he named it selenium from the Greek *selene*, for "moon."

Selenium turned out to have a number of uses. Since its electrical conductivity is affected by light, it became an important component of light meters. The same property led to its use as a toner in photocopiers. Today, the main use of selenium compounds is to impart a red color to glass, but nutritional research continues. There have been claims that selenium supplements can be helpful in treating cardiovascular and thyroid disease, as well as in avoiding cognitive decline. Looking at the evidence can leave you scratching your head. By the way, selenium sulfide is the active ingredient in some dandruff shampoos.

TOXICITY OF NICOTINE IS NOTHING TO SNEEZE AT

Sherlock Holmes was particularly fond of the gold snuffbox he had been given by the King of Bohemia for his assistance in preventing a scandal. He, like many Victorians, enjoyed sniffing the pulverized dried tobacco leaves, known as snuff, for the stimulant effect delivered by a swift hit of nicotine. That effect had been discovered by the Indigenous people of South America and was introduced into Europe by the Franciscan friar Ramon Pane, who accompanied Columbus on his second voyage to the New World in 1493.

By the sixteenth century, snuffing had become popular in Spain and Portugal, with medicinal effects being attached to the habit. Jean Nicot, French ambassador to Portugal, was particularly taken by snuff and described it as a panacea, a cure for all ailments. He introduced it to Catherine de' Medici, queen of France, who was so elated with the effect sniffing had on her headaches that she declared the tobacco plant would henceforth be called *Herba Regina* or "Queen of Herbs." That made snuff popular among the French nobility and its use quickly spread across Europe. Various scented snuffs were produced, and elaborate snuffboxes offered for sale. Napoleon, Marie Antoinette, Lord

Nelson, King Louis XIII, and Benjamin Disraeli all sniffed, but Czar Michael of Russia certainly did not. In 1643, he declared that anyone caught using snuff would have their nose removed. Seriously!

Diseases such as the plague and cholera were also instrumental in popularizing snuff. At the time, it was believed that many illnesses were caused by ill winds or "miasmas," and that pleasant odors such as those introduced by scented snuff could counter disease. Snuffing began to wane by the twentieth century, although a variety of scented and flavored snuffs are still produced. As a recognition of the history of snuff in Britain, a parliamentary snuffbox can be found at the door of the House of Commons. It has been there since 1694, when Parliament passed a resolution banning smoking in the House but tolerated snuff as an alternative. The last recorded use of snuff from the parliamentary box was in 1989.

Jean Nicot's contribution to spreading the fame of tobacco was recognized by Swedish botanist Carl Linnaeus, who introduced the botanical term *Nicotiana tabacum* for the plant. When its active ingredient was isolated by German chemists Posselt and Reimann in 1828, they named the oily liquid nicotine.

On entering the body, nicotine is rapidly distributed through the bloodstream and crosses the blood-brain barrier in seconds after inhalation. Snuff actually introduces more nicotine than smoked tobacco because heat destroys the compound. Nicotine is highly toxic, with about 40 to 60 milligrams being lethal for a human, usually through the paralysis of respiratory muscles. A cigarette, depending on the type, contains anywhere from 8 to 24 milligrams of nicotine, meaning that a pack of smokes easily has enough nicotine to kill a person. Of course, when tobacco is smoked, some of the nicotine is broken down, some is exhaled, and only a portion of what is inhaled makes it into the bloodstream. On average, about 1 milligram is absorbed from a cigarette, which is not enough to have a toxic effect but is enough to cause addiction. As far as snuff goes, only a pinch is sniffed, so toxicity is not an issue.

The toxicity of nicotine was recognized early on, as evidenced by herbalist and physician Nicholas Culpeper's recommendation in his classic 1681 book, *The English Physician*, to use tobacco juice to kill lice on children's heads. Nicotine would go on to be used as a commercial insecticide into the twenty-first century, when it was replaced by insecticides less toxic to mammals. The neonicotinoids were introduced as safer synthetic analogues of nicotine, but their use is now being questioned because of emerging evidence that they are harmful to bees.

Nicotine's potential as a poison appealed to Belgian Count Hippolyte Visart de Bocarme. In 1850, confident that at the time there were no means to detect nicotine in a body, he contrived to kill his brother-in-law to secure an inheritance by poisoning his food with nicotine he had distilled from tobacco leaves. He was successful, but officials were suspicious and approached chemistry professor Jean Servais Stas about the possibility of developing a method to detect nicotine in a corpse. The chemist rose to the challenge and managed with the use of appropriate solvents to extract nicotine, pioneering the first test to detect plant poison in human tissue.

Some 150 years later, Paul Curry thought he could carry out the perfect murder by injecting his wife with nicotine, motivated by getting his hands on her life insurance. Although a smart fellow, a *Jeopardy!* champion, he didn't reckon with the techniques that Stas had introduced. Tests revealed nicotine in Linda Curry's body and her husband was sentenced to life in prison.

Count de Bocarme wasn't quite as lucky. He was convicted and sentenced to death by guillotine. His last request was that the blade be very sharp so that his head would be severed with a single stroke. It was.

TREHALOSE AND *C. DIFFICILE*

The Exposition Universelle in 1855 in Paris was quite a spectacle. It drew over 5 million visitors, including Queen Victoria and Prince

Albert, the first British royals to set foot in France for over 400 years. Visitors were treated to exhibits of the crown jewels of France, the world's first sewing machine, carved caskets made of Goodyear's novel vulcanized rubber, and a Parisian inventor's 10-foot-high percolator that brewed 2,000 cups of coffee an hour. Crowds gathered around McCormick's reaper, a machine that revolutionized agriculture by making it possible to harvest grain more efficiently than men wielding scythes. Thanks to the reaper, food shortages were more unlikely since far more land could be now be farmed.

Also on display were items of medical interest, such as a sample of trehala, a powdered substance popular in the Islamic world with a reputation for relieving respiratory ailments. The name apparently was a corruption of "Teherani," meaning coming from Tehran, the capital of Persia. Trehala was made from the crushed cocoons of *Larinus maculatus*, a beetle that spends its life crawling on the leaves of a host plant. Whether trehala ever had any medicinal value is questionable, but it certainly has importance in the history of chemistry. It was the raw material from which in 1859 famed French chemist Marcellin Berthelot isolated trehalose, a simple sugar that would go on to have a variety of applications.

Long before Berthelot's experiments, trehala had made its mark under quite a different name. According to some biblical scholars, this insect residue was the "manna" that God had sent from heaven to feed the starving Israelites as they wandered through the desert after Moses had led them out of slavery in Egypt. The Bible describes manna as tasting like a wafer made with honey, which could well be a description of the taste of trehalose; it is composed of two molecules of glucose joined together and is indeed sweet. This sugar is widespread in nature, found in mushrooms, shellfish, algae, honey, yeast, and anything made from yeast such as wine, beer, and bread.

What makes trehalose interesting is its ability to preserve the cell structure in foods, especially after heating and freezing. It appears to replace the moisture in cells that is driven out by heat and thereby prevents the collapse of the network of proteins that is key to cell integrity. Undesirable changes in texture can also occur when cell structure is disrupted through the formation of crystals. Trehalose interferes with crystal formation in ice as well as in starch and therefore finds a use in ice cream, frozen foods, and baked goods.

Furthermore, foods and pharmaceuticals with added trehalose can be dehydrated and reconstituted with little damage. This protective effect is particularly handy when it comes to preserving the proteins that constitute vaccines, antibodies, and blood coagulation factors. With trehalose these can be stored in a dehydrated fashion requiring no refrigeration, ready to be reconstituted as needed, and solving the problem of a high percentage of vaccines in developing countries being wasted due to lack of refrigeration. Trehalose-dried blood is also a possibility and could be important in fighting critical blood shortages that are sometimes experienced.

Today trehalose is produced on a large scale from starch with the aid of bacterial enzymes and is used by the food and pharmaceutical industries as an additive, primarily because of its preservative properties. Extensive feeding studies involving both animals and humans have failed to reveal any adverse effects. Until now.

The heretofore sweet science of trehalose may be souring with the finding that some strains of *Clostridium difficile* bacteria have the ability to use trehalose as a nutrient. *C. diff* infections plague hospitals since these bacteria are resistant to many antibiotics and frolic when competing bacteria are killed off.

Since about 2000, *C. diff* infections have become not only more common, but more serious, puzzling scientists. Now there may be an explanation. A study published in the prestigious journal *Nature* in 2018 showed that some strains of *C. diff* can use trehalose as a nutrient, and when it is available, these strains outcompete others. Unfortunately, these are also strains that produce a higher level of the toxins that cause gastrointestinal distress. Interestingly, the increase in severity of the *C. diff* infections coincides with the discovery of an easy method to produce trehalose from corn starch and its introduction into the food supply.

At this point, the connection between trehalose and *C. diff* infections is best described as intriguing, but it is in need of further exploration. Calls for banning trehalose are premature, especially given that this sugar doesn't cause blood glucose to spike the way that other sugars do. However, the potential connection between trehalose and *C. diff* bacteria is an illustration of the complex interplay between our diets, our microbes, and our health. Even additives that have passed all regulatory safety requirements may have unintended consequences.

SEAWEED AND ITS SURPRISING SUBPLOTS

A story is told about a Japanese emperor and his party who sometime in the seventeenth century were lost in the mountains during a snowstorm and found a small inn where they were ceremoniously treated by the innkeeper, who offered them a jelly dish made with seaweed. Seaweed is a type of algae, defined as aquatic organisms capable of carrying out photosynthesis.

It seems the dish wasn't exactly to the emperor's liking and ended up being discarded. The night was cold, and in the morning the innkeeper noted that the seaweed concoction had frozen. When he passed by later, it had thawed and separated into a liquid and a solid layer. Draining the liquid left a crumbly solid that to the innkeeper's amazement could be used to reconstitute a gel by boiling in water and then cooling. This gelling agent came to be known as "agar-agar," or just "agar," from the Malay term for the red algae from which it is derived. Agar is widely used in Asian cuisine for jellied desserts and can also be found as a thickener in soups, jams, and ice cream. It is essentially a vegetable analogue of gelatin and is acceptable in vegan cuisine.

After the introduction of the germ theory of disease by Louis Pasteur in the nineteenth century, researchers began to take great interest in culturing bacteria in the lab. In 1882, German microbiologist Walther Hesse introduced agar as a medium for growing bacteria while working as an assistant in the lab of Robert Koch, who would go on to win the Nobel Prize for identifying the causative agents for tuberculosis, cholera, and anthrax. Supposedly Hesse got the idea after he had noted his wife's use of agar to thicken a soup that retained its consistency even after extensive boiling, which was not the case with gelatin-based thickeners. Gelatin had been previously used as a growth medium for bacterial cultures, but it tended to liquify as the temperature increased. In an agar medium, however, microbes could be cultured at higher temperatures, making it easier to determine which bacteria were susceptible to heat and which were not.

Agar's application in microbiology, and its introduction to the U.S. as a food-gelling agent, increased demand for the substance. Since it was mostly supplied by Japan, Western countries had to look to their native seaweeds for extracting agar when the Second World War broke out. These mostly turned out to be unsatisfactory, but they did yield other useful gelling agents such as the alginates, carrageenan, and furcelleran.

Today, commercial agar, composed mainly of a polysaccharide called agarose, is produced by a technique similar to the one that led to its original discovery, namely boiling then freezing seaweed. A number of novel uses include as an impression material in dentistry, modeling clay for children, fertilizer in organic farming, and as a diet aid. The latter is based on agar's ability to absorb water and leave the dieter with a feeling of fullness.

While agar is obtained from a specific species of red seaweed, other types of "macroalgae" also have commercial importance. A different red seaweed is being explored as an additive in cattle feed with a view towards reducing methane emissions. When cows digest their food, they belch large amounts of methane, a gas that makes up about 5 percent of total greenhouse gases. Bromoform, a naturally occurring component of some seaweeds, can reduce emission by interfering with the function of enzymes in the cow's rumen that lead to the production of methane.

What about seaweed as a food for humans? There are some 10,000 species of seaweed, with a diversity of flavor and nutritional properties. Although sometimes called a "superfood" by marketers, the vitamins and minerals found in seaweed are readily available from other plant products. The dried sheets of nori, used to make sushi, weigh so little that they are inconsequential in terms of nutrition. Some people enjoy snacking on seaweed wafers, which are low in calories and, surprisingly, sodium, making them preferable to chips or crackers. The taste of seaweed is largely due to naturally occurring glutamate, the same substance that people are concerned about when it is used as an additive (monosodium glutamate) in foods. Indeed, MSG was first isolated from seaweed in 1908 by Japanese biochemist Kikunae Ikeda.

While there appear to be no reports of adverse reaction to glutamate in seaweed, there is some concern about its iodine content. Iodine is essential for thyroid hormone formation, but too much can cause the same symptoms as deficiency, including goiter, which is an enlarged thyroid gland. Most people have no problem applying moderation to seaweed consumption.

Finally, powdered agar dissolved in hot water can be poured into a mold to form a thin layer of a solution that is then frozen, thawed and air-dried. The result is a film that is being explored as a possible biodegradable substitute for plastic packaging. You can even eat it.

THE POWER OF LYCOPODIUM

Would you like to make some money? Bet a friend that you can remove a coin from a glass of water with your hands without getting your hands wet! Gloves of course are not allowed. So, what is the secret? Lycopodium powder! What is that? A very fine dust consisting of the reproductive units, or spores, of a type of fir commonly known as club moss. A curious property of this powder is that it forms a smooth layer when sprinkled on top of water. The tiny individual particles have such an affinity for each other that when a finger is inserted into the water, the layer does not break but instead forms a protective coating around the finger. The coin can be thus picked up and the powder is shaken off to reveal completely dry fingers! Amazing!

Lycopodium powder has another interesting property. It burns in air with great vigor. There are two reasons for this. First, it is composed of highly combustible organic compounds such as sugars, starches, and tannins, and second, the powder is so fine that each particle can be surrounded by a large number of oxygen molecules that are necessary for combustion. If a little lycopodium powder is sprayed into a flame, it flares and burns with amazing speed. Sometimes it is sold in joke shops as Dragon Breath. But if lycopodium powder ignites in a closed container, the result is no joke! What we then have is an explosion!

The rapid combustion produces a great deal of carbon dioxide and water vapor, both gases that expand rapidly under the high temperatures produced. Pressure quickly builds, and the expanding gases, having no place to go, blow the container apart. This is a demonstration often carried out in chemistry classes to show how the rapid production of gases can lead to an explosion.

Nicéphore Niépce knew all about lycopodium explosions. That name may ring a bell since Niépce is usually credited with being the inventor of photography. Around 1816, he managed to capture pictures on paper coated with light-sensitive silver chloride using a camera obscura, basically a darkened box with an aperture on one face through which an image is projected onto the opposite inner surface. A decade before this, together with his brother Claude, Niépce built the world's first internal combustion engine. It didn't run on gasoline. The Pyreolophore, as it was known, ran on controlled explosions of lycopodium powder and was used to propel a boat!

Roughly twelve times a minute, lycopodium powder was injected into a stream of air that transported it into a combustion chamber where it was ignited by a smoldering fuse. The expanding hot gases produced by the explosion of the powder then powered machinery that sucked in water at the front of the boat and forced it out through a tailpipe at the back. The pulses of water pushed the boat forward. Lycopodium turned out to be very expensive and was replaced by pulverized coal, which in turn was eventually replaced with petroleum distillates. The Niépce brothers had produced the world's first fuel-injection engine!

It is not only lycopodium powder or coal dust that can cause an explosion. Any finely divided combustible material will do. For example, the Space Shuttle's booster engines used fine aluminum powder to create a controlled explosion that helped launch the vehicle into space. While controlled dust explosions can be useful, uncontrolled ones, such as in grain elevators, can be deadly. Grain particles can generate combustible dust as they are shifted around or moved along conveyor belts. When the right mixture, or in this case the wrong mixture, of dust and oxygen is arrived at, even a tiny spark can set off an explosion. The force of the explosion can be powerful enough to make concrete and steel grain elevators buckle like paper. A grain dust explosion in Louisiana in 1977 resulted in thirty-six deaths, an explosion of flax dust at a textile factory in China in 1987 killed fifty-eight people, and an explosion of colored starch powder released at a "music and color festival" in Taiwan in 2015 caused fifteen fatalities and almost 500 injuries.

There is yet another curious use of lycopodium powder. It is prescribed by homeopaths to treat people with "lycopodium characteristics." These are shyness, a lack of self-confidence, and "complaints that begin on the right side and move to the left." People with a lycopodium personality also supposedly get aggravated between 4 and 8 P.M.

The lycopodium patient, one homeopath tells us, is normally retiring and outwardly calm, but just like the normally inert lycopodium powder that explodes when thrown into a flame, is capable of sudden bursts of violent temper when things become "just too much" for them. The treatment for this condition is "Lycopodium 200C," which means that the powder is diluted with sugar to the extent that the final product does not contain a single molecule of lycopodium. Maybe I have a "lycopodium personality" because such bunk is enough to make me explode. And not only between 4 and 8 P.M.

BURGERS FROM PLANTS

I've always had a special respect for peas. That's because of the giant role they played in the advancement of science. It was through the crossbreeding of peas with different characteristics such as color, plant height, and pod shape that Austrian friar Gregor Mendel laid the foundations of the modern science of genetics. In experiments conducted between 1856 and 1863, he demonstrated how traits were passed on from one generation to another through invisible factors that were later identified as genes.

I don't know whether Mendel ate his peas, but I do know that many a parent has urged their offspring to eat the peas that accompanied the meat on their plate, with an eye towards increasing their vegetable consumption. Today, they may be encouraging them to eat their peas instead of the meat. Pea protein has become a hot item as the basic ingredient in plant-based burgers, the sales of which are skyrocketing due to concerns about the health and environmental consequences of animal agriculture.

Beyond Meat, an American company that produces plant-based meat substitutes, made a huge splash in financial markets when its stock tripled in three days after its initial offering. That was followed by soaring sales in supermarkets and fast-food restaurants that offered Beyond Meat burgers made with pea protein. Other ingredients include mung bean protein, canola oil, coconut oil, potato starch, methylcellulose, lemon juice, lecithin, and pomegranate fruit powder.

Beyond Meat's major competitor is the Impossible Burger, produced by Impossible Foods, a company founded in 2011 by Dr. Patrick Brown, who is no scientific slouch. He has both a medical degree and a PhD in biochemistry and is the recipient of many prestigious awards. In 2010, Brown left Stanford University, where he had been a professor, having become convinced that raising animals to produce food was an environmental disaster, citing greenhouse gas emissions, energy demands, and inefficient use of land as problems. He decided that the best way to reduce the environmental cost of raising animals was to offer a plant-based product that could compete with meat in look, scent, and taste.

The first question that had to be answered was what makes cooked meat look, smell, and taste the way that it does. A tough question because meat is an immense collage of numerous proteins, peptides, amino acids, fats, vitamins, minerals, steroids, amines, sugars, and nucleotides that engage in various reactions when exposed to heat. Cooked meat is therefore even more complex, being composed of hundreds and hundreds of compounds! Using sophisticated instrumental techniques, Dr. Brown and his team analyzed the volatile compounds released when meat was cooked, and their attention was drawn to heme, a breakdown product of the oxygen-carrying molecule hemoglobin, which was present in abundance. Brown suspected that it was a major player in the flavor game.

Thanks to his research background, Dr. Brown knew that some hemoglobin analogues could be found in plants. Leghemoglobin (from "legume" and the Greek *heme* for "blood"), for example, can be found in the roots of clover or soybeans, and it became a candidate for imparting meat flavor and color to a plant-based burger. The problem was that the compound could not be extracted from this source on a large scale. However, isolating the gene that codes for the formation of leghemoglobin and introducing it into the genome of a yeast cell was possible. When this yeast is combined with appropriate nutrients it dutifully grinds out leghemoglobin, which can be isolated and purified. Needless to say, this annoys anti-GMO activists, but of course leghemoglobin is the same substance whether it is produced in soybeans or genetically modified yeast cells. It is a nonissue.

The use of leghemoglobin for flavor and color was a breakthrough, but it still took about five years of experimentation to come up with a burger that had satisfactory texture, mouthfeel, and flavor. Eventually, soy protein, coconut oil, sunflower oil, potato protein, methylcellulose, yeast extract, dextrose, modified starch, salt, and various vitamins joined with leghemoglobin. In this case, environmental concerns are more relevant than nutrition since in terms of saturated fat, the Impossible Burger contains about the same amount (8 grams) as a hamburger but contains five times as much sodium! It would be hard to claim that the Impossible Burger is healthier, but it is more environmentally friendly. Same goes for the Beyond Meat burger.

People talk of health and the environment, but the decision about what to eat often comes down to taste. I haven't had the chance to taste the Impossible Burger, which is not yet available in Canada, but I did let A&W's Beyond Meat burger loose on my taste buds. And they didn't rebel. However, the rest of the body was not so keen on the 1,100 milligrams of sodium, most of which comes from the bun and condiments. The patty itself has only 390 milligrams, so next time maybe just the patty wrapped in lettuce? I'll have to see how the taste buds react.

CAN FLUORINATED COMPOUNDS SHRINK PENISES?

Edinburgh druggist John Scott was so taken by the achievements of Benjamin Franklin that in 1815 he established an award to be given annually to a "most deserving man or woman" whose invention has contributed in some outstanding way to the "comfort, welfare, and happiness" of mankind. The John Scott Award was to be administered by a committee in Philadelphia, the city where Franklin carried out most of his scientific work, including the possibly apocryphal kite-flying experiment. The awardees have included the likes of Marie Curie, Thomas Edison, Frederick Banting, and Jonas Salk.

At the 1951 award ceremony, all guests received a surprise gift. A muffin pan! But this was no ordinary muffin pan. It was coated with Teflon, the nonstick substance that had been discovered by Roy Plunkett, the DuPont chemist who was that year's John Scott Award honoree.

Teflon had been an accidental discovery. In 1938, DuPont had become concerned with the toxicity of ammonia and sulfur dioxide, refrigerants that were commonly used at the time. Plunkett's search for a replacement led to experimenting with tetrafluoroethylene, a gas he had made and stored in cylinders. One day he opened a cylinder and was mystified. No gas came out despite the cylinder having lost no weight!

Curiosity aroused, Plunkett cut the cylinder open and discovered a white powder that had some surprising properties. It was resistant to heat as well as to virtually every chemical he tried, including oil and water. But the substance's most impressive property was that nothing would stick to it. On further investigation, Plunkett realized that the small molecules of tetrafluoroethylene had joined together to form a polymer, later to be christened Teflon.

Determination of the molecular structure of Teflon revealed a chain of carbon atoms, each bonded to two fluorine atoms. It is this array of fluorines around the periphery of the molecule that turns out to be responsible for Teflon's nonstick property as well as its lack of reactivity. As soon as this was realized, researchers began to churn out a variety of fluorinated compounds, looking to produce stain and water-resistant materials. Numerous polyfluorinated alkyl substances (PFAS) were developed, finding uses in paper products, textiles, carpets, paints, medical equipment, plumbing tape, dental floss, food packaging, wire insulation, camping gear, and firefighting foams.

The use of these chemicals in firefighting foams was prompted by a fire aboard the aircraft carrier USS *Forrestal* in 1967 that killed 130 people when a rocket was accidentally launched into armed and fueled planes on the deck. The foam used was unable to spread over the burning fuel to smother it. This led to the development of foams that contain various polyfluorinated compounds that enhance the foam's fluidity. Subsequently equipment to dispense the novel foams was installed on military and civilian ships, airplanes, and in airports.

By the 1970s, some of the shine began to wear off the widely used polyfluorinated compounds with the discovery that they were environmentally persistent and were accumulating in the blood of occupationally exposed workers. Then in the 1990s, PFAS were being detected in the blood of the general population as well as in some drinking water supplies. By the turn of the century, scientists had linked exposure to some PFAS to a variety of health issues in animals, including reproductive problems and cancer.

Research revealed that the most worrisome compounds had eight or more carbons in their molecular structure and that shorter chain analogues would break down more quickly and were less likely to bioaccumulate. The industry switched to these, but their long-term safety has not been established.

In any case, due to their environmental persistence, exposure to the longer chain PFAS can still be a problem. This was demonstrated by an Italian study that found young men who had grown up in an area contaminated with these chemicals due to industrial production had significantly smaller penises, less mobile sperm, and a shorter anogenital distance than those who drank water with no PFAS. To learn why this should be so, the researchers studied the effect of these chemicals on cells in the laboratory and discovered that they can bind to testosterone receptors and block their activation. As a follow-up, they found that increased levels of PFAS in the blood correlated with higher blood levels of testosterone, meaning that testosterone was being prevented from entering cells, possibly explaining the smaller penis size.

This study generated articles that urged people to throw out their Teflon pans and featured headlines like "Something in Your Kitchen May Have Made Your Penis Smaller." While chemicals released during the manufacture of Teflon may be a concern, the pans themselves do not release any significant amount of polyfluorinated compounds into food. Still, the overall use of these compounds, in view of their potential hormone-disrupting activity, requires further scrutiny. There is, however, a difference between using these "forever compounds" to save lives, as may be the case with firefighting foams, and using them to render pizza boxes or fast-food wraps greaseproof.

FROM DRY CLEANING TO SARAN WRAP

It's a captivating story. Back in 1825, a careless maid in the employ of a Jean Baptiste Jolly knocked over a turpentine lamp, spilling the liquid onto a stained tablecloth. Apparently Jolly was a frugal man, since turpentine, a liquid obtained by the distillation of pine tree resin, was a cheap alternative to whale oil, the common fuel for lamps. While Jolly may have been thrifty, he was astute. He noted that the stains vanished where the kerosene had soaked through the fabric. And at that moment the dry cleaning industry was born! At least so goes the possibly apocryphal tale. What is known is that Jolly did open the first "dry cleaning" establishment under the name Teinturerie Jolly, using turpentine to bathe fabrics that would normally not stand up well to water. Dry cleaning is only "dry" in the sense that no water is used.

Although Jolly is commonly credited with the discovery, a patent for "dry scouring" was actually filed in the U.S. by Thomas Jennings four years before the "careless maid" episode. Unfortunately, the patent was destroyed in a fire, so the solvent Jennings used isn't known, but it was likely either turpentine or a petroleum distillate. Jennings, the first African-American to ever be granted a patent, became rich from his discovery and used his earnings to help fund the abolitionist movement.

The great scourge of the dry cleaning industry was the flammability of the solvents. This was such a great problem that the establishments had to be located away from populated areas. By the turn of the twentieth century, the search was on for solvents that would dissolve stains and were not flammable. Way back in the 1830s, Michael Faraday had synthesized a compound called perchloroethylene that turned out to be a useful non-flammable solvent but didn't generate much interest until the flammability issue in the dry cleaning industry stimulated research into the use of chlorinated solvents.

In 1933, Ralph Wiley was a college student working at the Dow Chemical Company in a lab where research on dry-cleaning solvents was being carried out. Wiley ranked low on the corporate ladder, and one of his jobs was cleaning glassware. One day he was confronted by a residue in a flask that defied all his cleaning efforts. No matter what he tried, a dark green, greasy film remained. He brought this to the attention of his boss, John Reilly, who was intrigued but was too involved in other projects at the time. Eventually when Wiley was hired by Dow as a chemist, the mysterious material in the flask project was resuscitated. Wiley and Reilly determined that molecules of vinylidene chloride, the chemical that was the focus of the dry-cleaning solvent research, had combined with each other to form long chains of polyvinylidene chloride (PVDC).

By 1943, the two chemists managed to work out a viable synthesis of this material and christened it "Saran," combining "Sarah" and "Ann," the first names of John Reilly's daughter and wife. Saran turned out to be useful as a protective layer against salty sea spray for fighter planes on aircraft carriers, as an insole for military boots, and as a stain-resistant coating on car upholstery. It was in 1953 that Saran hit the public spotlight after Dow had found a way to produce polyvinylidene chloride as a thin, clear film. "Saran Wrap" became an instant hit as a packaging material because it was impervious to oxygen and moisture, extending the shelf life of packaged food. Furthermore, when stretched, its coiled molecules would unravel and attempt to spring back to their original shape, resulting in the film's ability to "cling" to itself and to other surfaces.

In 2004, Saran Wrap, which had been acquired from Dow by S.C. Johnson & Son, underwent a dramatic change in formulation. Although the name was retained, the wrap was now made of polyethylene instead of polyvinylidene chloride. By the company's own admission, this was an inferior product with poorer barrier and cling properties. Why the change then?

Issues had arisen about the fate of plastics that contained chlorine, and PVDC had loads of it. When exposed to high temperatures, such as in an incinerator, chlorinated plastics were found to yield highly toxic dioxins, a problem that does not arise with polyethylene.

Polyethylene's greater permeability to gases is not necessarily a minus. For example, in the case of packaged meat, the passage of oxygen through the plastic allows for the formation of oxymyoglobin, which has the attractive red color that people associate with fresh meat. In the absence of oxygen, deoxymyoglobin has a dark purple color. While packaging meat for a few days in polyethylene is fine, this thin plastic is not appropriate for freezing meat since it will allow moisture to pass through, resulting in freezer burn. But polyethylene can be formulated as straight or branched chain polymers with different densities, strengths, and permeabilities. "Freezer bags" combine different types of polyethylene to prevent freezer burn.

It would be some sort of poetic justice if the plastic bags used by dry cleaners today were made of PVDC since that plastic was spawned by dry cleaning research. But like Saran, they are made of polyethylene. Be sure to recycle them.

CHLORINATED CHICKEN

The American poultry industry wants to put a chicken in every British pot. But consumers in the U.K are putting up a big squawk about the prospect of "chlorinated chicken" in their stew. They worry that this could happen once the U.K. disengages totally from the

European Union, which has banned American poultry since 1997. Once untethered, the British government will be free to make trade deals with the U.S., and American chickens may soon be crossing the pond. A bird-brained decision according to many Brits, who are cocksure the American birds will peck away at the safety of the British food supply. One could say they are crying fowl. But we won't.

The European ban never had anything to do with the safety of treating slaughtered chickens with chlorine, or any of the other chemicals, such as chlorine dioxide, peracetic acid, or cetylpyridinium chloride, that are used as disinfectants. The fact is that residues of these chemicals are not detectable. At issue was that the use of disinfectants might lead to poorer standards in all steps of processing because of the belief that the chemical rinse will take care of any bacteria that may be present. In Europe, measures about ventilation, lighting, and density in chicken houses are strictly enforced, while in the U.S., flock densities are allowed to be much higher, leading to lower production costs but a higher incidence of bacterial infection. There is also the risk that heavily soiled birds, a result of overcrowding, may not be sufficiently disinfected even if they pass through a chemical spray or bath.

That possibility is supported by a study from Southampton University in which microbiologists found that listeria and *Salmonella* bacteria remain active even after a chlorine wash. Although laboratory tests showed that the bacteria could not be cultured, suggesting that they had been inactivated, roundworms exposed to them died. There is also a question about the testing that is done on the assembly (or perhaps better said de-assembly) line, by American inspectors. Birds are randomly removed and placed into a bag with a solution that collects any remaining bacteria. This solution also contains chemicals that neutralize any disinfectants still present.

However, questions have been raised about whether the disinfectant chemicals are indeed effectively neutralized by the solution. If not, they will keep killing bacteria between the time the bird was removed from the line and the testing is completed, usually the next day. This means that the testing can indicate that the birds are more free of bacteria than they actually are. Of particular concern is cetylpyridinium chloride, which apparently is not readily neutralized during sampling and can produce false test results. The issue here isn't the safety of consumption, since proper cooking kills bacteria, but rather cross-contamination of other foods in the kitchen before the chicken is cooked. That's why there are recommendations to not wash poultry under running water, since the splashing can contaminate surfaces and foods in the vicinity.

What about the claim that "chlorinated chickens" are toxic? While it is true that chlorine can react with organic compounds to form potentially toxic organochlorides, such as chloroform, the presence of these in poultry flesh is negligible. Actually, they are found in far lower concentrations than in chlorinated drinking water. Furthermore, when chlorine dioxide is the disinfectant used, as is increasingly the case, there are no organochlorides formed.

While processing chemicals may not be an issue for consumers, there is concern that they can affect the health of workers in abattoirs, where concentration in the air may exceed safety levels. There have been various reports of workers suffering respiratory problems, especially after line speeds were increased. In some cases, 175 birds pass an inspection point per minute! This means that some visibly contaminated carcasses may not be picked up by inspectors, a concern that has resulted in several disinfectants being used instead of just one, and sometimes at higher concentrations with a consequent increase in aerosolized chemicals. Of particular concern are respiratory problems and burning eyes due to peracetic acid, which is formed when the disinfectants hydrogen peroxide and acetic acid are combined.

Unsurprisingly, the poultry industry disputes any problem with processing chemicals and has some backing from academic researchers. Professor John Marcy of the University of Arkansas's Poultry Science Department points out that the chemicals used are strictly regulated and that they reduce the risk of foodborne illness. However, the CDC reports that about 380 people in the U.S. die annually of foodborne *Salmonella* poisoning, while Public Health England reports no cases. Dr. Marcy also uses a rather curious argument when promoting the safety of peracetic acid, saying that it is a combination of hydrogen peroxide, the same kind you may have in your bathroom closet, and acetic acid, which is the acid found in household vinegar. That is true, but totally meaningless. Once hydrogen peroxide and acetic acid are combined, the peracetic acid formed has totally novel properties. You cannot tell anything about the safety of a substance based on the reagents from which it is formed. By such logic, one could say that chlorine gas is safe because it can be made by mixing household bleach with vinegar. That of course is not the case; chlorine gas is highly toxic and has actually been used as a chemical warfare agent.

Now that the U.K. has left the European Union, it will likely have to rely more on the U.S. as a trading partner, and the Americans will put on pressure to make the import of American chicken part of a trade deal. It will be interesting to see if the British attitude that equates "chlorinated chicken" with Chicken Little's squawking about the sky falling will change if supermarkets start crowing about the low price of American chicken.

QUACK QUACK

I have to tip my hat to the algorithms that Google and Amazon use to detect people's interests and then target them with specific ads. Since I often carry out searches that use terms like "alternative medicine," "naturopathy," and "water treatment," I am inundated with ads for products that seem to fit these categories. I thought it would be fun to make a collection of some of the more interesting products that were pitched to me over one week.

There were magnetic toe rings for weight loss, healing crystals from India, acupuncture-point-stimulating Chinese balls, and detoxifying

Tibetan singing bowls. But the one that really caught my attention was the Belly Button Tool, designed to "push the greatest button you will ever push." What is it? A wooden rod with a silicone tip that fits into the belly button (not sure what to do if you have an outie). How do you use it? "Rapid, gentle, in and out movements, going clockwise for 5 minutes when you wake up and 5 minutes before you go to sleep."

Why? Because this manipulation will lead to "weight loss, increased energy, relief from stress and anxiety, relief from pain, an increase in flexibility by 25 percent, and a raised immune system." A "loosened gut" to boot. How does tormenting your belly button in this fashion accomplish these wonders? According to the "Doctor of Detox" who markets the gizmo, it stimulates the vagus nerve, which is desirable since "vagus nerve dysfunction is responsible for numerous ailments." He should know because he is a "medical intuitive," a Doctor of Natural Medicine, and a Doctor of Humanitarian Services. Not sure why a finger would not work just as well.

This outstanding humanitarian is also concerned about the water we drink. We learn that "as water travels the earth it becomes symmetrically structured in a sturdy and elaborate geometric shape, reducing its surface tension, neutralizing toxins, and increasing its hydration power." Unfortunately, water treatment "wipes the memory of water clean." Luckily for us, this icon of science knows how to remedy the situation. He sells devices that can restructure the water. "Structured Water creates a greater flow of energy as it connects with life and when taken into our body, it greatly enhances our body's ability to rejuvenate and function in a more optimal manner." A portable Gold-Plated Structured Water device will set you back $1,495, but if you can forgo the gold, you can avail yourself of this miracle for $349.

Another ad prompted me to purchase Dr. Hidemitsu Hayashi's Original Hydrogen Rich Water Stick. This is to be immersed in water to hydrogenate it and "increase health and vitality." There is no mention of how this is accomplished. The stick turned out to contain bits of magnesium that can indeed react with water to produce hydrogen gas, although I did not note any bubbles when I followed the instructions. I should point out that hydrogen gas is virtually insoluble in water, and any that does dissolve would quickly outgas. Nevertheless, I did as I was told and consumed the "hydrogenated water" for a week. Not even a burp.

Next, I ordered the Kikar Portable Electric Hydrogen Water Ioniser. This promised not only to add hydrogen to water but to "convert it to the hydrogen anion which is a powerful antioxidant because it scavenges reactive oxygen species." The device turned out to be a bottle with electrodes at the bottom that when connected to electricity carried out electrolysis. This is a classic experiment in which water is broken down into oxygen and hydrogen. This time bubbles did indeed form, but the notion of converting hydrogen to anions through electrolysis is total nonsense. The literature with the product also claims that "hydrogen has been found to have anti-tumor effects," hoping to snare some desperate people.

By now I was getting kind of stressed, so I turned to another invention that promised relief, the Quartz Crystal Elixir Water Bottle. Drinking from it would "boost my spirits, gently neutralize negative vibrations, bring emotional calmness and relieve stress and anxiety." Surely this would work since it was backed by that shining star of science, Gwyneth Paltrow. The bottle contained an attractive large crystal, the properties of which were said to be transferred to the water "through molecular vibrations, charging it with energy, soothing our mind and emotions." The only emotion it generated in me was irritation.

The same can be said for PolarAid, a "revolutionary, easy-to-use, portable, handheld body tool that supports every aspect of health by harnessing the vital frequencies that naturally surround us at all times." True, it is easy to use. It is nothing but a plastic disc that looks like a coaster. It can even be used as such. We are told that a glass of water becomes energized when placed on it, and inserting it under a potted plant makes it thrive. For health, PolarAid can be used to "unblock blocked energy channels by holding it over chakra points." Once the channels are unblocked, health is ensured by sitting on PolarAid for thirty minutes a day. All this gives me indigestion. I think I may need a belly button massage.

ALPHA SPIN MAKES YOUR HEAD SPIN

I came upon the Alpha Spin device, if you can call it a device, in a roundabout way. I was looking into "Longevity Village" in China, so dubbed because of the unusually long life expectancies of its residents. Proportional to its population, the village has about five times as many centenarians as the rest of China. Located in Bama county, it has become a hot spot for "health tourists" who come hoping their ailments will respond to the magic of the environment, the same magic that allows the locals to have exceptional longevity.

Some believe the secret is to be found in the air of the area's caverns, naturally enriched with negative oxygen ions. One man claims he beat lung cancer through tai chi exercises in the cave and a diet of boiled pigeons and apples. Others hug giant boulders, convinced that this "geomagnetic therapy" is the key to health. However, most believe that the secret is to be found in the river that winds its way through Bama county. Tourists bathe in the river and consume endless varieties of "longevity water" that are sold everywhere.

It was while searching for research on Bama water that I made the acquaintance of Alpha Spin. Googling "Bama water," the following cropped up: "It is believed that the natural resonance of Alpha Spin is similar to that found in many water springs around the world, including Bama, an internationally recognized longevity village in China. Alpha Spin brings Bama to you by optimizing the natural frequency, stimulating vital life energy, and increasing harmony in body and mind." And down the rabbit hole I went.

I quickly learned that Alpha Spin is a "powerful holistic wellness tool" that "fully optimizes the body's molecular and cellular functions via resonance and the formation of a vortex that results in the expression of a quantum energy field." Furthermore, "the unification of the quantum energy field results in a metatron cube, the result of the interaction of spins allowing the pyramids to communicate by transferring information from quantum energy."

There was more about Alpha Spin's ability to generate hexagonal water clusters, transferring resonance frequencies through water, light, or air and neutralizing and harmonizing the harmful effects of electromagnetic frequencies. Wow! In all my years of wading through swamps of claptrap, I don't think I have come across anything to match the stew of random, garbled, meaningless words cooked up on behalf of Alpha Spin.

Alright, let's put the nonsensical rhetoric aside. What is Alpha Spin supposed to do? We are told that it can energize water and anyone who drinks said water. There is even a way offered to prove this, the "O ring test." The "O" is formed by placing your thumb and forefinger together. Then with your other hand, you try to pry the fingers apart and repeat after drinking the "energized" water. The "evidence" is that it is harder to pull the fingers apart the second time because you have been "energized." This is a well-known pseudoscientific ruse based upon expectations. If you believe that it will be harder to pull the fingers apart, then that will indeed often be the case. Of course, when such experiments are done in a blinded fashion, nobody can tell if the water has been "energized."

Alpha Spin also claims to extend the shelf life of fruits and vegetables, improve plant growth, reduce wrinkles, improve circulation, remove energy blockages, treat autism, increase the engine performance of cars, cure cross-eyes, stop jet lag, improve sleep, and of course, treat cancer. That is because, as everyone knows, "healthy cells have harmonic frequencies that rotate counterclockwise and cancer cells have clockwise spins." And amazingly, Alpha Spin changes the direction of the spin.

As you can imagine, I just had to have this wonder product. It was available on Amazon, so I went for it. Set our office back over $200, but hey, that is a small price for a miracle. I really had no idea what I was getting and waited eagerly for its arrival.

The attractive box said "German Technology" but was mailed from Indonesia. Inside was a little glass plate that looked like a coaster with a hole in the middle. It didn't buzz, didn't spin, had no flashing lights, didn't do anything. It was nothing but a glass disk. For the wondrous effects, you were supposed to hold it over a body part, or a glass of water or the hose when filling your car. As for establishing an effective quantum energy field to protect against electromagnetic radiation, I learned that I would have to invest another $600 because four disks need to be placed in the corners of the room in the shape of a quadrangle. Oy vey.

As far as Longevity Village goes, it seems that through natural selection, inhabitants are blessed with genes that code for a protein, apolipoprotein E, that plays a role in a number of important biological processes linked with health. And Alpha Spin? All it can do is make your head spin with nonsense.

FALSE CLAIMS AND CONSPIRACY THEORIES

It started innocently enough with a question from a reporter about the wondrous health claims being made on behalf of oil of oregano. I had dealt with queries about this oil before, essentially debunking the claims, but I thought it would be judicious to do a bit of research to see if any new information has come to light. Little did I dream that I would be embarking on a path that would unmask a popular champion of herbal remedies and reveal his alter ego as a proponent of the most outrageous conspiracy theories.

I had first looked into oregano oil a few years ago after listening to a radio interview with Dr. Cass Ingram on CJAD in Montreal. Since CJAD also hosts my show, I knew I would be getting some questions about the claims being made. Ingram was introduced as a "nutritional physician," whatever that means, and proceeded to describe how oil of oregano was the answer to sore throats, cough, congestion, and the flu. He went on to promote his book, *The Cure Is in the Cupboard*, basically a litany of the supposed benefits of oregano oil in the treatment of everything from allergies to zits. Dr. Ingram was well-spoken, sounded rational, and made some seductive arguments on behalf of "natural therapies." However, a literature search quickly revealed that the claims were exaggerated and not supported by clinical evidence. True, various extracts of oregano oil have antimicrobial effects in a test tube, but there is nothing exciting about that. Numerous plant-derived chemicals have such effects but then fail to show any value in human trials. After all, the body is not a giant test tube.

Having concluded that *The Cure Is in the Cupboard* was not trustworthy, I didn't pay much more attention to Ingram even though, much to my annoyance, he periodically reappeared on CJAD as a "doctor who has given answers and hope to millions through lectures on thousands of radio and TV shows." Nevertheless, since I was now looking to update my oregano oil knowledge, I thought it would be interesting to see if Ingram had added anything to his repertoire. It turns out that he had, but not about oregano.

A couple of curiosities immediately emerged. Sometimes his name was spelled Ingram and sometimes Igram. When I searched YouTube, I found a number of his videos. On one, using the name "Health Hunter," he promotes the benefits of wild Chaga. On others, he praises "black seed oil" as a miracle and promotes raw hemp oil as a wonder product. It is clear that he is firmly entrenched in the tenets of "natural cures," making wild, unsubstantiated claims. I guessed that he would be against vaccinations, and sure enough, I found a video of him spewing the anti-vaccine drivel. Just when I was ready to write him off as one of the horde of health cranks polluting legitimate science, I came across another video where he was no longer Dr. Cass Ingram or Igram, but was Dr. Kaasem Khaleel, discussing how Islam was misunderstood. This triple identity seemed bizarre and stimulated a deeper dive into the world of Ingram/Igram/Khaleel.

As it turns out, the man has a degree in osteopathic medicine, but in 1999 lost his Illinois medical license for "unprofessional, unethical, and dishonorable conduct," apparently charging a patient inappropriately for nutritional supplements. Some of the supplements he recommends in his books are sold by the North American Herb and Spice Company, which was taken to task by the Federal Trade Commission in the U.S. for making bogus claims about oregano oil. The company eventually agreed to pay $2.5 million to settle the charges. Ingram has also had issues with the Internal Revenue Service for nonpayment of taxes. But these are minor travesties when compared with his putrid writings as Kaasem Khaleel, or "Dr. K," another alias.

In his book *Wrongly Blamed: The Real Facts Behind 9/11 and the London Bombings*, he claims that international Jewry was behind the terrorist acts. The website nodisinfo.com, which was registered to Khaleel but is no longer online, professed that the Boston Marathon bombers were framed by the Mossad, the Israeli Intelligence Service and that the Sandy Hook school shooting was a hoax. When Khaleel guested on a broadcast of *The Realist Report*, his appearance was promoted on a page that also featured the line, "The Holocaust is the biggest and most persistent lie in history" and had sidebar quotes from Hitler and Goebbels.

It is rare to find a person whose decrepit blather about natural medicines takes a back seat to his other corrosive assertions, but such is the case for the character I first encountered as Dr. Cass Ingram. It seems another encounter is not in the cards because, in spite of quite an effort, I am unable to find his whereabouts. I would love to ask him and his alter ego a few questions.

As far as oil of oregano goes, there is actually something new. It seems the oil effectively repels a type of cockroach. Fitting that Khaleel/Ingram/Igram and cockroaches are in some way connected.

TURNING THE TABLES ON THE TABLE TURNERS

"What a weak, credulous, superstitious, ridiculous world ours is, as far as concerns the mind of man. How full of inconsistencies, contradictions, and absurdities it is." Those are not the words of some commentator bemoaning the lack of current critical thinking, although they well could be. They were uttered in 1853 by Michael Faraday, one of the greatest scientists who ever lived. Faraday's remarks were prompted by the craze of "table turning" that was sweeping across Europe. Participants at séances sat around a table with their hands placed on top waiting for spirits to demonstrate their presence by causing the table to turn. And it often it did! But as Faraday would go on to show, spirits were not involved.

By the time Faraday turned his attention to séances, he had already invented the electric motor, the electric generator, the transformer, and the process of electrolysis that led to the isolation of a number of elements. Indeed, it is no overstatement that he changed the world. And his popular public lectures at London's Royal Institution introduced people to the science of that changing world.

The birth of spiritualism, a movement based on the belief that the spirits of the dead exist and have the ability and inclination to communicate with the living, is often traced to the home of Kate and Margaret Fox in Hydesville, New York. It was there in 1848 that the sisters claimed to have contacted the spirit of a murdered peddler whose body had been found in the house. The communication was not verbal, but was in the form of rapping sounds audible to anyone present. People flocked to hear the messages that supposedly came from the murdered man in the other world. Actually, they were quite of this world. The girls had developed a remarkable ability to crack their toes! One could call them the original rappers! Incidentally, there never was any record of a murdered peddler.

From these bizarre beginnings, the spiritualist movement spread far and wide. Believers began to frequent sittings known as séances at which they attempted to communicate with the dead under the guidance of a medium. In the dim light of a séance, the spirits would often signal their presence by the movement of small objects, the playing of instruments without musicians, and the tilting and levitation of tables. Fraud was often involved, with mediums using magic tricks to convince the gullible that they had made contact with the spirits. This annoyed magicians who used the same effects for entertainment but always made clear that no supernatural phenomena were involved. And naturally, scientists like Michael Faraday who were well-versed in the laws of nature became interested in these supposedly paranormal events.

Faraday became curious enough to attend some séances and witnessed first-hand the movement of tables without any evidence of trickery. If it wasn't chicanery or the handiwork of spirits, in which he did not believe, then what caused the movement? Some scientists suggested that the motion was due to electrical or magnetic forces generated by the sitters, but Faraday demonstrated that this was not the case by covering the table with a variety of insulating materials through which electrical or magnetic forces could not pass. The table still frolicked.

Having dispensed with the possibility of human emissions, Faraday suggested that the sitters were subconsciously exerting a force with their hands, a phenomenon that would eventually be referred to as the "ideomotor response." Simply put, this occurs when a thought brings about a muscular response without the awareness of the subject. In the particular case of table turning, the séance participants hoped so much that the spirits would appear that their own muscles made it happen. But how to prove this?

Faraday devised a system of cardboards attached to each other and to the table by rubber cement in such a way that the top board on which the hands rested could move sideways before that motion was translated to the table. Sure enough, he was able to demonstrate to the satisfaction of the participants that it was their inadvertent hand motion that tipped the table. The sitters were honest, intelligent people who deceived themselves by engaging in muscular activity that was consistent with their expectations.

As he stated, Faraday had managed to "turn the tables on the table turners," but nevertheless spiritualism persisted. He resolved to continue the battle through education, the object of which was "to train the mind to ascertain the sequence of a particular conclusion from certain premises, to detect a fallacy, to correct undue generalization, to prevent the growth of mistakes in reasoning." Restating that goal today is well warranted.

As far as spirits go, we cannot prove that they do not exist. But if they do, it is curious that they have nothing better to do than move furniture about.

FINALE: ALWAYS WEAR UNDERWEAR

Tempus fugit. Yes, I took Latin in high school, which really does date me. It means time flies. Does it ever! It is hard to believe that twenty-two years have passed since I wrote the first book in this series, *Radar, Hula Hoops and Playful Pigs*. Even harder to believe is that forty-one years have passed since I answered a call on a dial telephone from CJAD radio host Helen Gougeon. "Would you like to comment on the controversy in the *Gazette* article?" she asked. And that conversation would lead to four decades of demystifying science for the public on the radio. Yup, I have now been on CJAD since 1980, on what I like to call the longest-running radio show on chemistry in the history of the world. (Of course, it may be the only radio show on chemistry in the history of the world.)

Here we need a little history. It was back in 1980 that organizers of the UNESCO pavilion at the "Man and His World" exhibition, a descendant of Expo 67, approached my colleagues Ariel Fenster and David Harpp and me to see if we would be interested in putting on a series of chemistry shows. They had heard that we had been exciting students with our lectures and chemical demonstrations. Sounded like a great opportunity for demystifying chemistry for the public, and we jumped. Over two summers, thousands would be attracted to our performances, including Prime Minister Pierre Trudeau, who brought along a very young Justin.

One of our featured demonstrations was the production of some polyurethane foam that involved mixing two reagents in a cup and generating, within minutes, a mountain of foam that hardened into a mushroom-shaped blob. It was a neat demo. We had a lot of fun with it until a fly fell into the ointment one morning. I remember it well.

I picked up my morning *Gazette* and began to glance through it in the usual fashion. Ted Blackman's "city column" immediately got my attention. It was all about our chemical escapades! Ted described how in spite of the great anxiety about urea-formaldehyde foam insulation, some chemists were brewing the stuff in public and were singing its praises. That got me more than a little hot under the collar. True, there was concern at the time about urea-formaldehyde, an insulating material that can release toxic formaldehyde if improperly applied. But we were not dealing with urea-formaldehyde! We were demonstrating the properties of polyurethane, a distinctly different material. The only common feature was that these were both foams!

By nine o'clock that morning I had delivered a letter to Ted, typewritten in those days, along with a large egg formulated out of polyurethane, which I suggested he hang around his neck for penance. After all, he had laid a large egg by not appreciating the difference between urea-formaldehyde and polyurethane. Much to his credit, Ted followed up with a mea culpa retraction, explaining that he had leaped to an inappropriate conclusion, having skipped too many chemistry classes in high school.

I was satisfied and thought the case to be closed. That's when I got the call from Helen Gougeon asking if I would like to comment on this controversy, which of course was really a non-controversy. She must have liked the way I explained the matter because I was soon asked to be a regular on CJAD. To underline just how long ago that was, one of my first memories of working on radio was editing audiotape by cutting and splicing! The radio show would eventually lead to an invitation to write The Right Chemistry column for the *Gazette*. I recall submitting the first one on a floppy disk!

Over the years there have been questions galore, ranging from the serious to the amusing. A lady visiting her sister in the Caribbean found that while her sister's bed was overrun with ants, hers was free of the creatures. Could it have anything to do with her sister being a diabetic? Possibly. At one time physicians used to diagnose diabetes by tasting a patient's urine to see if it was sweet. And ants are known to go for sweets. Ditto fruit flies. A gentleman queried whether wine in which some fruit flies had drowned was safe to drink. It is. That question presented an idea for fruit fly control. Just leave a few glasses of wine around and wait till the flies drink themselves to death. It works.

There certainly have been questions I was unable to answer. Where does one go to get a goldfish autopsied if there is a suspicion it has been poisoned? What is the best way to remove the green color from an emu egg before dyeing it? Can a gentleman's lack of success on his honeymoon night have anything to do with having just consumed twelve bananas? I didn't dare ask about the motivation for the banana frenzy.

And last year I was asked about the COVID-denier who vowed never to wear a face mask or underwear because "things gotta breathe." Time to invoke my Latin knowledge. Semper ubi sub ubi. And masks too.

INDEX